AIR MOBILITY

AIR MOBILITY

The Development of a Doctrine

Christopher C.S. Cheng

Foreword by Lieutenant General Ellis D. Parker

PRAEGER

Westport, Connecticut
London

Library of Congress Cataloging-in-Publication Data

Cheng, Christopher C. S.
Air mobility : the development of a doctrine / Christopher C. S. Cheng ; foreword by Ellis D. Parker.
p. cm.
Includes bibliographical references and index.
ISBN 0-275-94721-1 (alk. paper)
1. United States. Army—Aviation—History. 2. Airmobile operations (Military science)—History. I. Title.
UG633.C495 1994
355'.166'09—dc20 93-5853

British Library Cataloguing in Publication Data is available.

Library of Congress Catalog Card Number: 93-5853
ISBN: 0-275-94721-1

First published in 1994

Praeger Publishers, 88 Post Road West, Westport, CT 06881
An imprint of Greenwood Publishing Group, Inc.

Printed in the United States of America

The paper used in this book complies with the Permanent Paper Standard issued by the National Information Standards Organization (Z39.48-1984).

10 9 8 7 6 5 4 3 2 1

To Debbie, my parents, and George

CONTENTS

Contents

TABLES

FOREWORD

Historians generally agree on the birth of American air power and its evolution is very well documented. However, very little scholarly work has been accomplished on the birth of Army Aviation. Most recognize the birth was not without difficulty and indeed the maturation of Army Aviation as a full-fledged combat arms branch continues today. This book completes a rather detailed survey of events, issues and key players that shaped Army Aviation and air mobility from its beginning in 1942 through 1965.

The author is to be commended for a very accurate accounting of our early history. It was developed from a blend of detailed research into the archives for documentation followed up by interviews relating to the documentation and finally a detailed analysis of what was discovered and collected. This approach enabled Dr Cheng to produce a descriptive piece of work with an analytical flavor not always present when we review our history lessons.

The book is not without limitations. Time and resource constraints prevented research into the National Archives and the Center for Military History. And as already mentioned, it does not proceed beyond 1965. Hopefully, Dr Cheng or someone else with his diligence and style will continue the saga, for the men and women of Army Aviation are making history every day.

The value of the book is twofold: First, Dr Cheng provides a wealth of detail on aircraft developments, including the years of work to form efficient organizations with capabilities for the full gamut of military operations. Second, he presents a comprehensive portrait of dedicated, visionary people struggling to provide the America's Army a potent combat multiplier, which eventually emerged as the Aviation Branch—full partner in the Army's Combined Arms Team.

Ellis D. Parker, Lieutenant General,
United States Army (retired)
Chief of Army Aviation Branch, 1984–1989

Preface

There is a paucity of literature on Army aviation history. (Unless otherwise specified, the terms "Army" and "Army aviation" refer specifically to the US Army and US Army aviation.) Part of the problem is due to the inaccessibility of source material. No comprehensive and accurate account of Army aviation history exists today. Existing accounts do not adequately explain the growth of Army aviation and air mobility. Most of the available literature, with a few exceptions, lacks analytical content or contains rather shallow analysis at best. Almost all works are lacking in historical details or do not cover the necessary period to give a proper sense of Army aviation development. Some even have more serious shortfalls such as historical inaccuracy. The primary objective of this book is therefore to give a more accurate historical account of Army aviation, and to analyse and explain the process of a relatively unknown Army innovation.

Modern Army aviation began on 6 June 1942, when the War Department approved organic aviation for the Field Artillery. Since this is such a basic prerequisite for later developments, and this move was not without controversy, this book begins from the pre–World War II period. Air mobility, evolved from artillery observation, began to develop in the Army just before the Korean war. The term "air mobility" originally defined just the air movement of military forces on the battlefield. It eventually encapsulated the idea of a tactical combat unit using the air medium to execute most of its functions. Full implementation of air mobility did not occur until 1965, when the Department of Defense created the first air mobile division. However, the period before 1960 was extremely important in developing the necessary prerequisites. This included developing the basic doctrine, organization, and tactics, developing and acquiring the proper aircraft and equipment, and conducting the necessary training. This book stops at 1965 because the creation of an air mobile division can represent the fulfillment of an innovation.

The majority of available works on Army aviation and air mobility assume that civilian intervention in 1962 was the key factor in the innovation, ignoring the important progress achieved before this intervention. Only recently have some writers begun to acknowledge the importance of the fifties. Indeed, senior Army officials took an active interest in Army aviation and air mobility. Their pro–Army aviation stance resulted in the significant progress made. Moreover, the changing international security environment was ultimately more important in determining air mobility developments than were civilian officials. This book offers an alternate analysis of how Army aviation and air mobility developed.

This book, however, also has limitations. Because of time and financial constraints, the research did not cover all potential sources of archives, such as the National Archives (both in Washington DC and Suitland) and the US Army Center of Military History in Washington DC. Therefore, this work does not purport to offer a comprehensive or even a definitive history of Army aviation and the air mobility innovation. Moreover, it does not cover the period beyond 1965, even though the evolution of Army aviation and air mobility continues. Indeed, one can argue that Army aviation did not become fully institutionalized until the early eighties, when the Army created a separate aviation branch.

I would like to thank the following people for the enormous help they provided. Without them, this book would have been even more difficult to complete. At the Army Aviation Center, Fort Rucker, Alabama: William J. Hayes, Public Affairs Office; Steve Maxham (Director) and Regina Burns (Librarian), Aviation Museum; Dr John Kitchens (Aviation Branch Command Historian) and Lieutenant Arthur Price, Command Historian's office; Beverly Hall (Chief Librarian) and Beverly McMaster (Reference Librarian), Aviation Center Technical Library. A special thanks must go to David Burrough, for opening an "Aladdin's cave" by pointing me towards the Military History Institute. At the Military History Institute: John Slonaker, chief historian; Mr Randy Rakers, security manager; Kathleen Gildersleeve, library assistant; Sandy Bauriedl, reference librarian; Pam Cheney, archives technician; Dr Richard J. Sommers, chief archivist-historian; David A. Keough, assistant archivist-historian; Judy Meck, interlibrary loan technician; Laura Popielski, receptionist; Louis Arnold-Friend and Dennis Vetock, historian. At the Air University Library: Robert Lane, Director, and Gene Johnson, Security Officer; and Allan J. Rod, Deputy Director, Public Affairs Office at the Air University. At the USAF Historical Research Center (Reference Division): Archie DiFante, Marvin Fisher, Dick Gamma, Sandy Smith, Essie Roberts, Frances Rawlins, SMSgt Gary McDaniels, SSgt Ed Gaines, Capt. George Cully (Enquiries), David Friday. All photographs are courtesy of the Army Aviation Museum.

I am indebted to Dr Phil Sabin for his advice and editing on the earlier version of this study.

Acronyms and Abbreviations

AA	*American Aviation*
AAAA	Army Aviation Association of America
AAF	Army Air Forces
ACR	Aerial Combat Reconnaissance
ACS	Assistant Chief of Staff
ACTIV	Army Concept Team in Vietnam
AFT	*Air Force Times*
AGF	Army Ground Forces
AH	*American Helicopter*
AID	*Army Information Digest*
ANAFJ	*Army Navy Air Force Journal*
ANAFJ&R	*Army Navy Air Force Journal and Register*
ANAFR	*Army-Navy-Air Force Register*
APC	Armored Personnel Carrier
ARST	Aerial Reconnaissance and Security Troop
AT	*Army Times*
AUSA	Association of the United States Army

AW	*Aviation Week*
CAA	Civil Aeronautics Administration
CDEC	Combat Developments Experimentation Center
CG	Commanding General
CONARC	Continental Army Command
CRD	Chief of Research and Development
DCSOPS	Deputy Chief of Staff for Military Operations
DOD/DoD	Department of Defense
FECOM	Far East Command
FY	Fiscal Year
JAAFAR	Joint Army and Air Force Adjustment Regulations
JCS	Joint Chiefs of Staff
MAAG	Military Assistance Advisory Group
MOMAR	Modern Mobile Army
MR	*Military Review*
MRRP	Materiel Requirements Review Panel
NDTJ	*National Defense Transportation Journal*
OCAFF	Office, Chief of Army Field Forces (became CONARC in February 1955)
OCT	Office of the Chief of Transportation
ODCSLOG	Office, Deputy Chief of Staff for Logistics
ODCSOPS	Office, Deputy Chief of Staff for Military Operations
ODCSPER	Office, Deputy Chief of Staff for Personnel
OSD	Office of the Secretary of Defense
PEMA	Procurement of Equipment and Missiles, Army
PENTANA	Pentagonal Atomic-Nonatomic Army
PPBS	Planning-Programming-Budgeting System

ROAD	Reorganization Objective Army Division
ROCAD	Reorganization Of the Combat Armored Division
ROCID	Reorganization Of the Combat Infantry Division
ROTAD	Reorganization Of The Airborne Division
STOL	Short Take-Off and Landing
STRAC	Strategic Army Corps
STRICOM	Strike Command
TAC	Tactical Air Command
TC	Transportation Corps
TEAM	Test and Evaluation of Air Mobility
TSMC	Transportation Supply and Maintenance Command
USAAD	*US Army Aviation Digest*
USAREUR	United States Army, Europe
VTOL	Vertical Take-Off and Landing

I. The Early Years

1

PRECURSOR: THE CREATION OF ORGANIC ARMY AVIATION

On 6 June 1942 the United States War Department created an organic aviation specifically for the Field Artillery. Its mission was to provide observation and direction for artillery fire. This occurred because the aerial artillery observation support provided by the then Army Air Corps fell short of the desired performance. Ground force officers, especially in the Field Artillery, argued that a cheap solution was readily available to rectify this growing gap. This involved the Field Artillery using its own simple-to-operate commercial light aircraft for such missions. The Air Corps and even some elements in the ground forces, however, opposed the idea. The Field Artillery's arguments won the day because there was an alliance of important personalities behind it, as well as because of the urgent nature of the problem. Top military and civilian officials in the War Department eventually heard and accepted the arguments to rectify the problem with observation aviation.

During World War II organic aviation operated very successfully in the artillery observation mission as well as missions that other segments of the Army called upon it to perform. The War Department eventually approved organic aviation for several other ground arms as well in 1945. Its success secured its survival in the postwar military reorganization. Congress passed the National Security Act in 1947 to create the Air Force and to unify the three services under a new umbrella organization called the Department of Defense. The act contained a general clause that provided high-level political and legal endorsement for aviation organic to the Army, or Army aviation. This event was fundamental to the future of Army aviation. The vague wording of the clause was to have the most far-reaching effect on air mobility development in the Army. Without such a vague endorsement, Army aviation would not have been able to develop and expand as it did.

ORGANIC AIRCRAFT FOR THE FIELD ARTILLERY

Aircraft Suitability and Command Arrangement

The Field Artillery's call to operate its own organic aircraft was the culmination of the poor quality and quantity of observation aviation provided by the Air Corps. The problem of quality concerned not only the type of aircraft, but also the command arrangements.

The problem of aircraft suitability originated from the increased weight and performance of Air Corps observation aircraft since World War I. The increasing weight meant that such observation aircraft required long prepared runways, rather than being able to take off and land on unprepared field strips close to the front. The increasing speed and altitude performance of observation aircraft meant that they were becoming increasingly unsuitable for artillery observation and fire adjustment missions. Such missions required flight at slow speeds and low levels.

This state of affairs existed during the middle to late thirties because of a combination of factors. First, the Air Corps term "observation" referred to five missions, which it derived from World War I experience: aerial reconnaissance over tactical areas; artillery fire control; aerial photography in hostile areas; contact-patrol flights with infantry units to keep commanders informed of positions held by advancing or retiring troops during a period of movement; rapid transportation of staff officers, messengers, and the like, for liaison purposes. The Air Corps dogmatically envisaged a single airplane for performing all five observation missions. Its ideal observation airplane was one that had a high rate of climb and was fast and maneuverable. It also needed armament and armor for protection against small-arms fire and hostile pursuit aircraft, and a slow landing speed.[1]

This may have been possible in the early aviation era when differences between classes of aircraft were marginal. However, aircraft performance began to differ widely soon after World War I. The high performance required for tactical reconnaissance in the enemy rear areas conflicted with what a liaison or artillery observation mission required. The Air Corps tendency was to develop observation aircraft with higher performance, thus sacrificing their ability to perform artillery fire control and liaison missions. The O-47, the standard Air Corps observation model from 1937 until Pearl Harbor, epitomized this development. The growth in aircraft complexity and maintenance requirements also demanded the use of airfields with better facilities. As a result, during the thirties heavier observation airplanes became sited at airfields further and further away from the units they supported.

The inability of observation airplanes to collocate with artillery headquarters seriously affected their working relationship and introduced a deteriorating command and organization problem. Moreover, there was a tendency for the Air Corps to group together squadrons that previously operated separately. This was because increasing numbers of aircraft operated from the same airfields, since airfields with adequate facilities were scarce. This created a problem with observation aviation doctrine.[2] In 1936 the War Department modified the earlier doctrine of assigning one observation squadron per division, to one observation group of four squadrons assigned to each corps. Theoretically each division still had one observation squadron in support, but in practice this procedure did not work to the satisfaction of the Field Artillery. Moreover, in reality there were not even enough aircraft to support this doctrine.[3]

The failure to differentiate between the competing and incompatible requirements within the observation mission was in part due to the Air Corps drawing erroneous lessons from its World War I experience. At that time numerous airfields were available close to the relatively static front. The Air Corps failed to realize that, in future, airfields might not be available close to tactical headquarters, especially ones on the move.[4] What the ground forces required for artillery observation and liaison, then, were light aircraft with a low-speed envelope. This would give maximum maneuverability and the ability to take off and land with minimal ground roll from short unprepared rough fields.

By 1938 the Air Corps began to realize its mistake and set about differentiating between the manifold observation missions. The catalyst for this change stemmed from European armies that had already recognized the potential of slow-flying light aircraft and were introducing them into service. One in particular, the German Fieseler (or Fiesler) Storch, so impressed the Air Corps that it soon recognized the airplane's tactical potential. As a result, in 1938 the Air Corps took steps to develop and procure from scratch a slow-flying aircraft purpose-designed and purpose-built to military specifications. It ran a competition for these airplanes with new high-lift devices such as slots and flaps. The Air Corps let the contracts in September 1939, but a lack of funds and the slow process of aircraft development and procurement introduced delays. The result was too little too late. The Air Corps did not receive its first short-range liaison observation airplane, the O-49, until September 1940. Even then the number of these military light airplanes entering service was pitifully insufficient. This delay ultimately was to help prompt a decision to procure commercial rather than military models. In the meantime, by October 1940 the Air Corps was planning to drop the pure observation type and use full combat airplanes like pursuit or bombardment types mixed with light liaison types.[5]

It seems that obtaining an appropriate observation aircraft did not concern the ground forces until 1940, when the Field Artillery had to develop arguments for organic aviation. For example, in 1937 the ground forces had a chance to help determine the type of observation airplane needed, but their response was less than helpful. Moreover, because of the cost and delay in obtaining the O-49 military light airplane, in November 1939 the Air Corps itself planned to make up the shortfall in numbers by procuring commercial models. At that time the Field Artillery dismissed the commercial light airplane, stating that it was good only for service test and that only a purpose-built aircraft was acceptable. What was of immediate concern to the ground forces was the pitiful level of observation aviation support. This deficiency surfaced during the 1940 maneuvers.[6]

Quantitative Shortfall

The catalyst that led ground officers to feel the aviation shortfall more acutely grew from the war in Europe and the large-scale army maneuvers in 1940 and 1941. The 1940 maneuvers, especially by the Third Army earlier in the year, showed up the pitiful level of aviation support. The delayed in-service and full operational date of the O-49 military airplane gave no hope to the ground arms that this situation would improve soon. On the other hand, production capacity for civilian light airplanes was plentiful, and manufacturing time was comparatively short. The ground arms therefore began demanding to operate commercially available light aircraft to solve the shortfall in aviation support. This call grew after unofficial contacts developed between ground force officers in the field and civilian light aircraft manufacturers, beginning in the summer of 1940. These contacts grew in intensity as flying salesmen flew to numerous Army bases to demonstrate their aircraft. Many senior officers were targets of these demonstrations. Even though the airplanes were short of the ideal, the commercially available light airplane had all the basic attributes necessary to deploy alongside field headquarters and small units. Light aircraft manufacturers themselves were eager to develop such contacts, as they had earlier tried unsuccessfully to interest the Air Corps in using their airplanes for flight training.[7]

The origin of the inadequate level of observation aviation came directly from the poor level of Air Corps readiness as a whole before World War II. This did not improve even after hostilities in Europe and the expansion program for the Air Corps began. Historians commonly put the inadequacies down to one factor: the Air Corps preoccupation with strategic air warfare. Some historians believed that this focus led to a general decline in tactical aviation, particularly observation aviation.[8] Whether this was the case or not, the Field Artillery began to feel that the best way to secure adequate support aviation was to own it. The Infantry, Cavalry, and the Armored Force also shared this sentiment.[9] As a result, the Field Artillery proposed the organic aviation idea to the War Department in July 1940 and in January 1941, but without success.

By spring 1941 the ground arms fully appreciated what commercial light airplanes could do and how inadequate the Air Corps supporting aviation was. Contacts between the manufacturers and the ground forces developed to such an extent that eventually civilian aircraft participated in Army maneuvers. In mid-1941 the ground arms were preparing to use a 12-aircraft "squadron" formed from various commercial manufacturers' airplanes for that year's maneuvers. Civilian pilots who were representatives of the manufacturers flew these airplanes. The Secretary of War informed General Henry (Hap) Arnold in June 1941 of the ground forces' plan to use these airplanes during the Second Army's June maneuvers in Tennessee. This gravely concerned General Arnold, Chief of the Air Corps. However, there was nothing he could do, as he realized that the Air Corps lacked the aircraft to fill the ground forces' needs. The Army repeated this arrangement in July and August with the Third Army maneuvers in Texas and Louisiana. By the end of July, the ground forces considered the light planes such a success that they requested permission to procure 20 airplanes. Though the War Department turned down this request, Assistant Secretary of War for Air Robert Lovett suggested paying the manufacturers for their expenses. The squadron continued on this rental basis to participate in other Army maneuvers.[10]

In the meantime, the Assistant Secretary of War for Air also began to feel that the purpose-built light airplanes for the Air Corps were not cost-effective. They cost around two-thirds of a high-performance observation airplane, while taking 6,000 hours to construct compared to 300 hours for a commercial light airplane. In 1939 a contract for 100 YO-49 cost the Air Corps $1,853,451.69. Meanwhile, the average price of a commercial light airplane had dropped from $5,000 each in 1930, to $3,500 in 1931, and eventually to $2,400 (with radio).[11] To Robert Lovett the purchase of the O-49 was an extravagant use of funds and facilities, especially during a period of massive Army Air Forces expansion. (The Air Corps became the Army Air Forces on 20 June 1941.) The earlier unofficial manufacturers' trials at maneuvers convinced him of the adequacy of commercial models. Therefore on 14 August 1941 he asked the Chief of the Army Air Forces to procure 12 commercial light aircraft for service test.[12] Low cost was therefore one important catalyst that led to mass introduction of commercial light airplanes.

The 1941 maneuvers also totally convinced the ground arms of the benefits of using commercial light airplanes, then widely known as puddle jumpers or Grasshoppers. They called for a massive expansion of supporting aviation assets possible only with these airplanes. In September 1941 General Lesley McNair (the Commanding General, Army Ground Forces) requested the procurement of 200 commercial light airplanes, which he raised to 500 by October. In November even the War Department felt that there was sufficient merit in the commercial airplane to use it as a substitute. It also felt that they would be the least detriment to the production of combat aircraft. The Air Corps acted on this advice and began negotiations for 617 commercial light airplanes. When the War Department held a "puddle jumper" conference in Washington DC on 14 January 1942, the Field Artillery alone laid down a requirement for 2,750 aircraft. The total number of short-range liaison observation aircraft requested for all the ground arms came to 4,000. The Air Corps agreed to supply this number. By February 1942 it increased the order for civilian light airplanes by 1,000, to 1,600 airplanes. Six months later the Air Corps increased the order by another 1,960 aircraft.[13]

In the same September as McNair's request, the poor levels of Air Corps cooperation in providing sufficient air units for the 1941 training maneuvers also became apparent. This prompted Major General Robert Danford (the Chief of Field Artillery) to propose organic aviation for the Field Artillery yet again on 8 October 1941. On 4 November 1941 the War Department again turned down this proposal.[14]

Support from Top Officials

While the appropriate airplanes were entering service, possessing these airplanes was a necessary but not sufficient condition for closer or better coordination between the aerial observer and ground commander. The original Air Corps plan in April 1940 was to assign only three military-model slow-flying airplanes to each observation squadron. However, the rapid growth in the demand for light airplanes meant a change of plans. In October 1941 the Air Corps proposed that a light observation squadron with 12 light liaison-type and 6 pursuit-type observation airplanes support each infantry division. It also proposed that a medium squadron with 6 liaison-type, 6 bombardment-type, and 6 pursuit-type observation airplanes support each cavalry and armored division and each corps. Each army was also to get two medium squadrons. The War Department approved this program on 10 December 1941.[15]

Despite the Air Corps plan to reassign observation airplanes to the division level, the Field Artillery's aspiration for operating organic aviation remained. It eventually managed to get approval for this. Perhaps it was because the Air Corps was unlikely to contemplate assigning light airplanes to individual artillery battalions. The basis of this opposition would be its aversion to subservient command arrangements and its preference for a centralized approach. Even so, War Department approval for organic aviation was unlikely even if there were such calls from General Danford, the Field Artillery, and others. This was especially true while there was significant opposition to the idea.[16]

The Air Corps was naturally against the idea of organic aviation owned by the Field Artillery. This would mean the Air Corps losing control over something that originally belonged to it. It alone was powerful enough to oppose the Field Artillery. In terms of personal stature and influence, the Commanding General of the Army Air Forces, General Hap Arnold, stood well above the Chief of Field Artillery, Major General Robert Danford. Furthermore, the Commanding General of the Army Ground Forces, General Lesley McNair, was also an initial opponent of the organic aviation idea. Even without these opponents, the War Department was unlikely to approve of organic aviation. First, the Air Corps was not an independent service and was still under the nominal control of the Army. It was more likely for the War Department to feel that the problem of observation aviation was reparable under this regime. Second, the War Department feared that to allow the Field Artillery to operate organic aviation would eventually lead to half-a-dozen separate air elements for all the other arms and services. Such opposition should have put paid to whatever arguments the proponents of organic aviation put forward.[17]

The approval for organic aviation was forthcoming because proponents solicited support from top military and civilian officials in the War Department. In this way the proponents were able to push the idea through.

Civilian light aircraft manufacturers not only developed extensive contacts with ground force officers in the field, they also generated intense lobbying aimed at the top civilian officials of the War Department. They also searched for sympathetic officials able to exert pressure. For example, the commercial airplane manufacturers' representative in Washington, John Morgan, had a longstanding personal friendship with Robert Lovett, the Assistant Secretary of War for Air. Lovett kept Morgan informed of light aircraft developments within the Army. Morgan also knew other prominent top civilian officials. He organized a congressional flying club in Washington and arranged for John McCloy (an Assistant Secretary of War) to take flying lessons. Both McCloy and Lovett offered practical suggestions to Morgan for his endeavor. Lovett was a key active participant on the issue of organic aviation. He was very much aware of the feud between the Air Corps and the ground forces over the inadequate level of air support. It was Lovett who suggested renting the commercial airplanes for the Third Army maneuvers in August 1941. The reason he suggested renting the planes instead of procuring them was that it did not require Air Corps approval. Moreover, it was Lovett who initiated the procurement of the first 12 commercial airplanes with his 14 August 1941 memorandum.[18]

Major General Robert Danford also approached top officials to build up support for organic aviation. After the War Department again turned down his proposals on 4 November 1941, he turned to two former artillery officers he knew in high office. First was the Secretary of War himself, Henry Stimson, who was in the same camp as Danford during World War I. Stimson sent him to see his Assistant Secretary of War, John McCloy. Danford described both Stimson and McCloy as very sympathetic to his call for an organic air observation post. McCloy became a strong and active proponent of organic aviation for the Field Artillery. He agreed to the Field Artillery conducting a service test, despite strong reservations from the Army Air Forces.[19]

As a result, the issue of organic aviation surfaced again. This time it did not originate from the ground forces, but from the Army Chief of Staff himself. On 3 December 1941 General George Marshall instructed War Department G-3 to issue a directive to test organic artillery aviation. The Assistant Chief of Staff, G-3, War Department, in turn made an informal telephone request to the Chief of Field Artillery to solicit brief general proposals for implementing this order. The Field Artillery submitted such a service test proposal on 5 December 1941. The Chief of Staff approved on 8 December and instructed the Adjutant General to issue the necessary instructions to the Army Ground Forces and the Army Air Forces. This occurred on 10 December, three days after Pearl Harbor.[20] On 15 January 1942 the training of Field Artillery personnel to carry out the service test started, using 28 YO-59 (L-4) borrowed from the Army Air Forces. After a period of training lasting until 28 February 1942, field tests began in earnest. The director concluded that the tests were successful, and sent a report to the Army Ground Forces.[21]

The timing of this report's submission may have been deliberate, made to coincide with the absence of General McNair. Brigadier General Mark Clark, Deputy to General McNair and Chief of Staff of Headquarters Army Ground Forces, concurred with the findings and submitted them to the War Department in his absence. The tests may have impressed General Clark previously during an inspection.[22] (What he subsequently did for Army aviation puts him firmly in the camp of organic aviation proponents.) The War Department approved the Field Artillery ownership of very light aircraft on 6 June 1942 without the concurrence of the Army Air Forces. The airplane's sole mission was for observation and direction of artillery fire. Each artillery battalion, and brigade or group headquarters, had two airplanes, two pilots, and a mechanic. Thus each infantry division had a total of ten artillery observation airplanes, while each armored division had six (later eight).[23]

SUCCESS IN COMBAT

The War Department decision limited the assignment of organic aviation to the Field Artillery and not to any other ground arm, even though they had also called for their own. This was because the Field Artillery had the most urgent and clearly identified need compared to others. Nevertheless, the needs of the other ground arms soon became apparent when ground forces began fighting. The rest of the division called on the artillery airplanes to perform missions other than artillery fire control. The battlefield success of the civilian light airplanes led the Army Ground Forces (AGF) to start calling for a formal extension to other ground units. It also led some ground elements to call for airplanes with performance better than the existing L-2, L-3, or L-4.

This success and the expansion requests initially had two effects on the Army Air Forces (AAF). First, it tried intermittently to absorb or recapture control over organic aviation. However, organic aviation's success, along with continued high-level patronage, ensured its bureaucratic survival. Second, the AAF tried to block any expansion beyond the existing user base (Field Artillery) or aircraft type (two-place commercial light airplane with a 65 horsepower engine). Part of its strategy to head off expansion demands was the formation of liaison squadrons, operating mostly L-5 airplanes with a 180 horsepower engine. It assigned one to each field army to supplement the divisional organic aviation and contain it within the division level. Even so, the field demand meant an *ipso facto* expansion of organic aviation, with it performing a broad mission for a broad user base. By the end of the war other Army ground elements became approved users of organic aircraft of the light, commercially available type.

Competition from the Army Air Forces

In January 1940 the Air Corps accepted the requirement for a short-range liaison observation airplane and it began to define the tactical functions of "observation" in detail. The formation of AAF liaison squadrons slowly evolved after this. The AAF was to use this type of aircraft for the following missions: a courier and messenger carrier; a convenient transport for unit commanders and staff officers; a command observation airplane suited to use in reconnoitering routes, camp sites, and airdrome locations; a plane to check camouflage; to perform conventional infantry and artillery missions of liaison and fire control. On 1 July 1942, the AAF again changed the composition of observation squadrons because of problems in the light squadron. It standardized both the light and medium squadrons into a squadron with six pursuit types, six bombardment types, and nine liaison types. (In April 1942 the AAF redesignated short-range liaison observation simply as liaison.) Despite the new structure, the low priority of observation aviation other than liaison was causing serious problems. The AAF on one occasion in October 1942 considered separating the observation and liaison functions and forming a consolidated liaison flight within the division. The lack of observation aircraft other than the liaison types was the problem.[24]

Since the lack of high-performance observation aircraft persisted, on 9 February 1943 a board convened to examine the subject of observation aviation. It agreed with the idea of withdrawing liaison aviation from observation squadrons and making it available to support ground units. The AAF accepted the recommendation to homogenize previously mixed squadrons, with the term "observation" changed to "reconnaissance." On 2 April 1943 the War Department authorized two reconnaissance squadrons (fighter), one reconnaissance squadron (bombardment), and one liaison squadron with 16 aircraft for each reconnaissance group. In May 1943 the War Department authorized the AAF proposal to organize one or more liaison flights. While one liaison detachment entered combat in the China-Burma-India theater in spring 1943, the majority of liaison squadrons were not ready for shipment overseas until early 1944. By this time the number of aircraft in the squadron, predominantly equipped with the L-5, had increased to 32. One squadron supported one army, with each corps receiving a flight of eight airplanes in support. By the end of the war over two dozen such squadrons were in the various theaters.[25]

On 11 July 1944 the War Department approved a broader range of missions assigned to the liaison squadrons. These included the following: messenger and courier service; transport and ferry service for ground force personnel and equipment; visual reconnaissance; light photographic reconnaissance and other limited photographic missions; column control; check upon passive air defense measures, such as camouflage, cover, concealment, dispersion and dummy installations; artillery adjustment, when required to supplement organic Field Artillery liaison aviation; limited air evacuation, when required in emergencies.[26] The liaison squadrons were also very successful on the battlefield. For example, during December 1944 the 25th AAF Liaison Squadron resupplied the entire 11th Airborne Division on the island of Leyte. A report on the operations of the 25th Liaison Squadron said the squadron performed 529 supply drop missions because "it was impossible to supply them by any other means." The report continued:

> The light planes proved more advantageous than C-47s for drops because the percentage of equipment and supplies recovered was materially higher due to their ability to drop more accurately and at a lower air speed. They were able to pinpoint their drops.[27]

The activities of the AAF liaison squadrons illustrate that the expansion of organic Field Artillery airplanes did not occur in a vacuum. This made its expansion all the more impressive.

Expansion of Organic Aviation Missions and Operators

There was only a brief time between the approval of organic aviation (6 June 1942) and the initial entry into ground combat in North Africa (8 November). This period proved too short for assimilating organic aircraft properly into ground units. Before the Field Artillery had fully developed their utilization, commanders used the aircraft freely as the aerial equivalent of the ubiquitous jeep. This demand persisted even after the Field Artillery was exploiting them fully in fire control missions. Almost every type of ground combat unit found a use for them and borrowed them from artillery units whenever it could to meet some vital need in combat. As a result the mission of organic artillery aircraft grew unofficially to include the following: courier and liaison service; visual and photographic reconnaissance by ground reconnaissance units (mechanized cavalry), infantry, armor, tank destroyer, and engineer units; column control by swiftly advancing armor; emergency resupply; speedy evacuation of wounded from the front lines; close-in bombing; relay communications between front-line ground troops and supporting fighters or bombers and for indicating targets to them, known as "Horsefly" missions; directing naval gunfire; performing aerial surveys.[28]

While organic aviation performed many different missions, this did not mean neglect of the artillery mission. On the contrary, artillery observation became one of its most successful aspect. The Field Artillery's airplanes controlled an overwhelming majority of observed artillery fire missions. Artillery fire directed by aerial observers was so effective that often the mere fact an artillery airplane was airborne was enough to silence enemy guns.[29]

Despite the existence of AAF liaison squadrons and AAF attempts to block any extension or improvement of organic aviation, the AGF eventually succeeded. This was not due to the failure of the liaison squadrons, but because organic aviation was more successful and overshadowed them. It is difficult to discern the secret of organic Army aviation's success compared with the AAF liaison squadrons (see Table 1-1). The difficulty is in separating the fact that the Field Artillery's airplanes were organic and operated by Army personnel, as opposed to the centralized liaison squadrons operated by the AAF. Daily association may be the determining factor. Perhaps ground force personnel understood ground force problems better. It is impossible to determine whether organic aviation operated by the AAF or centralized Field Artillery aviation would have been as successful or not.

Table 1-1
A Comparison of Field Artillery and AAF Liaison Aircraft Organization and Control

Organic Army Aircraft	Attached AAF Liaison Squadrons
Advantages:	*Advantages:*
• The airplanes were more readily available to the using unit. • It simplified command and control of the aircraft because the pilots were officers of the using unit. • It simplified procurement of personnel to fly and maintain the aircraft. • Liaison pilots trained in the basic tactics of the using arm appreciated better the problems of the ground unit, felt they were members of a team, and became imbued with the unit's "esprit de corps." • It expedited supply and maintenance of items other than Air Corps items. • It simplified administrative problems such as maintenance of records and proper provision of messing, billeting, and medical supplies and facilities.	• It relieved ground commanders of supply, maintenance, and other administrative responsibilities, except for certain functions such as messing and quartering. • It assured the availability of air force information and technical experience when communications to the liaison squadron's parent air force unit were available. • Facilities existed within the air forces to train pilots and to equip and supply liaison units. • Centralized control permitted attachment to ground force units as prescribed by theater or army group headquarters, thereby assuring the economical use of the aircraft.
Disadvantages:	*Disadvantages:*
• It might not effect maximum economic use of the aircraft through failure to coordinate and combine missions. • Air Force facilities such as weather information, traffic and flight regulations, and services offered at airfields might become less readily available. • The necessity for establishment of additional training facilities to meet the increased air training demand on the part of the ground forces might be uneconomical. • The maintenance burden would require the establishment of additional units and complication of supply problems.	• The ground unit served suffered because: (1) The temporary nature of attachment made it difficult for the commander to rely on the continued use of the aircraft. (2) Personnel not trained by the ground forces were not of maximum value to ground force units. (3) Divided command responsibility between air and ground force commanders made complete unit control by either force impossible. • The attached liaison squadron suffered from the divided responsibility. (1) Separation from their own higher headquarters complicated supply, maintenance, and administration. (2) The nature of the attachment and the lack of training with the ground force units resulted in a morale problem or lack of squadron "esprit de corps." (3) Liaison personnel feeling "orphaned" from their own unit adds further complication.

At first, the War Department steadfastly maintained the status quo between organic aviation and the liaison squadrons. It turned down both the AGF requests for extensions and improvements, and the AAF bids to recapture organic aviation.[30] Eventually the War Department began to change its position over the issue of extension. Instrumental in this change was the involvement of civilian officials and a study done by the AAF that gave a favorable view of organic Army aviation.

John McCloy not only maintained an active interest in organic aviation, he also suggested its extension.[31] Robert Lovett also played a key role in trying to persuade the AAF and perhaps the War Department to accept the extension of organic aviation. This began when on 10 October 1944 the AAF circulated a memorandum that renounced the claim to organic aviation. As it transpired, the Chief of Air Staff who signed the memorandum had not coordinated it with other sections of the air staff. Indeed, the air staff attempted to recall the document, with many still holding anti–organic Army aviation sentiments. The memorandum explicitly stated that the Assistant Secretary of War for Air coordinated the recommendations.[32] Thus it is likely that Lovett himself initiated this move to make the AAF renounce organic aviation. Maybe the weight of combat experience convinced Lovett of the merits of organic aviation. Perhaps he also sought to reduce the contentions between the Army ground and air arms.

Between January and April 1945, the AAF instigated a major study of its postwar policy toward organic Army aviation. This study took a favorable view of the extensions the ground forces were asking for. This study was well received in some parts of the air staff. However, an AAF board convened in June to consider this issue rejected the notion of expansion.[33] The AAF resisted until July 1945, when General Jacob Devers, Commanding General, Army Ground Forces, and Lieutenant General Ira Eaker, Deputy Commander, Army Air Forces, came to an agreement. A policy memorandum issued by General Eaker on 3 August cemented this agreement. This memorandum contained guidelines that became AAF policy toward organic Army aviation after the war. Primarily, such aircraft must be "of a type similar to those suitable for [air observation post] use." Moreover, the AAF would favorably consider requests for organic aircraft if and only if:

> a. The aircraft will be put to a sustained use, and one which exploits its maximum capabilities.

> b. Separation of the aircraft from the mass of air power, considered in terms of aircraft, will not seriously reduce that power.
> c. The function for which it is proposed to use the aircraft is not one which, if performed by the requesting arm, will duplicate the capabilities of Air Force units and equipment without a compensating increase in the national capability for waging efficient war.
> d. No concommitant [*sic*] necessity will arise for separate and extensive airdrome, depot, maintenance or training facilities.[34]

On 9 August the War Department finally approved the extension, but while in the process of establishing the extended program the war ended on 14 August. Nevertheless this gave six additional airplanes to each infantry, airborne, and mountain division, nine to each armored division, and seven to each cavalry division. It also assigned two airplanes to each nondivisional cavalry squadron, separate tank battalion, tank destroyer battalion, cavalry group, and tank destroyer group. Each separate engineer battalion also got one airplane. In terms of aircraft improvements, in early 1945 the War Department decided to supply the AGF with the L-5 as interim to supplement the L-4. Ultimately, the L-4X (later redesignated the L-14) was to replace both, but its production was pending. In the end the war terminated before any L-14 entered service.[35]

POSTWAR POLITICAL ENDORSEMENT

The National Security Act of 1947

The process of unifying the War Department and the Department of the Navy began as early as April 1944 with congressional hearings on postwar military policy. Yet the process did not proceed in earnest until after the end of the war. Unification finally came into being after President Truman proposed a draft bill to Congress on 26 February 1946. The House of Representatives and Senate approved a revised version of the President's proposals on 25 July 1947 after lengthy hearings and debate. Organic Army aviation was not a point of contention, since it already existed and operated with success during World War II. Indeed, it would be difficult to abolish with such a record. The President approved the compromise version of the draft bill on 26 July 1947 and it became Public Law 253, 80th Congress, widely known as the National Security Act of 1947.[36] Section 205 (e) of the National Security Act of 1947 stated:

> In general the United States Army, within the Department of the Army, shall include land combat and service forces and such aviation and water transport as may be organic therein. It shall be organized, trained, and equipped primarily for prompt and sustained combat incident to operations on land.[37]

This law did not further elaborate on the functions or composition of the "aviation as may be organic therein." Subsequent documents issued by the executive stated the functions of the Army in much the same general manner as the National Security Act. This included Executive Order 9877 (26 July 1947) and the "Functions of the Armed Forces and the Joint Chiefs of Staff," or the so-called "Function Papers" (21 April 1948).[38] This did not change when Congress modified the National Security Act in 1949. The President passed the legislation and it became Public Law 216, 81st Congress, on 10 August 1949. This failure to clarify the details of Army aviation had two implications. First, it allowed Army aviation a high degree of flexibility. Second, it effectively left the responsibility to the Army and Air Force to find an appropriate delineation of aviation missions. This was without doubt the cause of intense conflict between the Army and the Air Force. The Army would be attempting to maximize its aviation functions as much as possible, while the Air Force would be trying to minimize them as much as possible. While the two services were able to arrive at a compromise, it did not last long.

Interservice Agreements

The first postwar attempt at delineating aviation functions occurred in the 1946 War Department Equipment Board Report. The purpose of this board, chaired by General Stilwell and known as the Stilwell Board, was to outline a dynamic research and development plan for the postwar period.[39] The field manual FM 20-100 *Army Ground Forces Light Aviation* published on 9 September 1947 also listed the missions of organic aviation. Both recommended that the Army Ground Forces use organic light aircraft to accomplish missions performed during the war (see Table 1-2). The field manual noted that the Air Force remained responsible for providing liaison squadrons for Army use, much in the same fashion as in World War II and for the same missions. The Army used organic aircraft at division level or below, while higher headquarters used Air Force liaison squadrons. Each army and each theater or task force headquarters had one liaison squadron attached. Each squadron still had four flights of eight aircraft, although it might have more than four flights. It might also have one or more utility cargo or helicopter flights if required.[40]

The two services reached their first interservice agreement entitled "Army–Air Force Agreement as to the initial implementation of the National Security Act of 1947, 15 September 1947." This did not clarify the limits of organic Army aviation and the responsibilities of the Air Force. It merely stipulated that the Air Force must still organize, equip, and operate liaison squadrons for use by the Army. The cross-procurement arrangement was to continue. It also stated that the Army would provide surface transport and the Air Force would provide air transport for both services. On 17 September 1947 the two services presented the agreement for the approval of James Forrestal, the Secretary of Defense.[41]

On 20 May 1949 the two services reached another agreement on the question of organic Army aviation. Published as the Joint Army and Air Force Adjustment Regulations or JAAFAR 5-10-1, it had the title *Combat, Joint Operations, Etc.: Employment of Aircraft for Performance of Certain Missions*. This document explicitly stated the limits and functions of organic Army aviation. Fixed wing aircraft could not exceed a weight limit of 2,500 pounds. Rotary wing aircraft, which the Army had already acquired by 1949, had a weight limit of "3,500 to 4,000" pounds. Specific functions outlined combined those recommended and proposed earlier. However, specification of weight limits was a new development. During the war the War Department did not allow the AGF to go beyond the two-place commercial light airplane. Now it was possible for the Army to acquire aircraft with more seats as long as they were within the weight limits and satisfied the conditions contained in the next agreement. The Air Force retained the responsibility of providing liaison squadrons for various services and missions (see Table 1-2).[42]

Table 1-2
A Comparison of Proposed and Agreed Army Aviation and Air Force Functions, 1946–1949

Stilwell Board Report 29 May 1946	FM 20-100 9 September 1947	Army Proposal Late 1947	JAAFAR 5-10-1 20 May 1949
For the Army:	For the Army:	For the Army:	For the Army:
• Close-in observation. • Battlefield reconnaissance, both visual and electronic. • Adjustment of fire of short-range field artillery, rockets, and guided missiles.	• Maintaining continuous aerial surveillance on enemy forward areas. • Locating appropriate targets. • Adjusting fire. • Obtaining information on hostile and friendly forces.	• Maintaining aerial surveillance of enemy forward areas. • Locating appropriate targets. • Adjusting fire. • Obtaining information on hostile and defensive forces.	• Maintaining aerial surveillance of enemy forward areas for the purpose of– (1) Locating appropriate targets. (2) Adjusting fire. (3) Obtaining information on hostile and enemy defense forces.
	• Aerial route reconnaissance.	• Aerial route reconnaissance.	• Aerial route reconnaissance.
• Control and liaison.	• Control of march columns.	• Control of march columns.	• Control of march columns.
	• Camouflage inspection flights.	• Camouflage inspection of ground force areas and installations.	• Camouflage inspections of ground force areas and installations.
• Local courier and messenger service.	• Messenger service. • Courier service.		• Local messenger and courier service.
• Limited aerial photography.	• Aerial photography.		• Limited front line aerial photography.
• Individual and group transportation, evacuation of casualties, and emergency supply within the combat area.	• Aerial supply in cases of emergency. • Aerial evacuation in cases of emergency. • Radio relay.		• Emergency aerial evacuation. • Emergency wire laying. • Limited aerial resupply.
For the Army Air Forces:	For the Army Air Forces:	For the Air Force:	For the Air Force:
• Air support missions involving the use of high performance aircraft, in general, should remain a responsibility of the Army Air Forces.	• Courier. • Light transport. • Visual & light photo reconnaissance. • Column control. • Camouflage inspection. • Limited aerial supply. • Air evacuation of casualties.	• Courier service. • Messenger service. • Aerial evacuation. • Aerial supply. • Aerial photography.	• Courier service. • Messenger service. • Aerial evacuation. • Aerial supply. • Aerial photography. • Aerial wire laying.

On 23 March 1950 the Army and the Air Force agreed on a further set of guidelines. This agreement, published as Army Regulation 700-50 and Air Force Regulation 65-70, assigned prime responsibility for Army aircraft administration to the Air Force. The Army could only procure either Air Force–developed aircraft, commercial aircraft certified by the CAA,[43] or modified commercial aircraft certified by the CAA and approved by the Air Force. The Air Force was responsible for preparing specifications, conducting research and development at the Army's request, and making type classifications in coordination with the Army. It also handled the purchase, acceptance, and inspection of aircraft and associated equipment. Moreover, the Air Force was responsible for a host of other activities related to testing, supply, and maintenance of aircraft and their associated equipment. The Army retained the authority to determine the qualitative and quantitative requirements and funded Air Force activities performed on its behalf.[44]

Conflict Between the Army and the Air Force

The souring of interservice relations began in 1946 when the Constabulary Force Headquarters in Europe organized a flight detachment using its own Army aircraft assets. The AAF had disbanded, as far as is known, all its liaison squadrons there, despite the increasing aviation requirement. The headquarters organized the provisional Constabulary Flight Detachment with 20 L-5 and 2 C-45 airplanes, and submitted the table of organization and equipment to the War Department for approval. Despite the controversial nature of the request, during late 1947 the Army also submitted this flight detachment proposal to the Air Force. It also proposed to abrogate and replace the 15 September 1947 agreement with one that mirrored the 9 September Army aviation field manual. The Army wanted to redefine organic aviation as consisting of "light aircraft" rather than liaison airplanes, which were "commercial low performance aircraft of the 'Piper Cub' type" specified on 6 June 1942. It also asked that individual aircraft or those "in groups of such size as to permit performance of assigned missions" be allowed for assignment or attachment to Army organizations. Moreover, it wanted to make official a list of expanded functions, though it still asked the Air Force to provide light aircraft to perform certain missions (see Table 1-2).[45]

Elements of the Air Force asked to study the problem to formulate recommendations for dealing with the Army strenuously opposed both these moves. Tactical Air Command (TAC) was especially hard-line and demanded a return of all organic Army aviation to the Air Force. The Air University accepted that the Army had a requirement for liaison-type aircraft, but it opposed the operation of such aviation above division level and the proposed Army agreement. The Air Force opposed the formation of the flight detachment because it was then not organic aviation, and it argued that the Army had no authorization to organize aviation units. It also opposed the new proposal that allowed this. The main Air Force concern here was the eventual emergence of an "Army Air Arm." TAC contended that if the Army could form aviation units, it "would permit them to seek their own transport aircraft as being indispensable in the utilization of airborne troops. Once Troop Carrier was lost, it was conceivable that reconnaissance, fighter, and bomber aircraft would follow. Tactical Air Command recognized this as a dangerous possibility."[46] Thus the Air Force immediately reactivated the 14th Liaison Squadron and stationed it at Heidelberg, Germany, in support of the Constabulary. This move did not succeed because the Air Force could not equip the squadron as the Army had all the liaison-type aircraft in Europe. The Air Force also opposed the Army's proposal to change the description of liaison aircraft into light aircraft. As the Air University pointed out, the term "light aircraft" was so general that it could mean even light bomber.[47] The 1949 agreement, however, allayed Air Force concern for the time being.

NOTES

1 Captain Irving B. Holley, Jr, Army Air Force Historical Studies: No. 44, "Evolution of the Liaison-Type Airplane, 1917–1944," AAF Historical Office, Headquarters, Army Air Forces, April 1946, pp. 1–2.

2 (1) Brigadier General William W. Ford, "Forty Years of Army Aviation, Part 1: Grasshoppers," *USAAD*, June 1982, pp. 4–5. (2) Richard Tierney, *The Army Aviation Story* (Northport, Alabama: Colonial Press, 1963), pp. 42–43.

3 (1) Dr Robert F. Futrell, USAF Historical Studies: No. 24, "Command of Observation Aviation: A Study in Control of Tactical Airpower," USAF Historical Division, Research Studies Institute, Air University, Maxwell Air Force Base, Alabama, September 1956, pp. 1–2. (2) Maurer Maurer, ed., *Air Force Combat Units of World War II* (USAF Historical Division, Air University, Department of the Air Force, 1961), pp. 5–7.

4 Holley, pp. 5–6, 113–14.

5 (1) Lieutenant Colonel Andrew Ten Eyck, *Jeeps in the Sky: The Story of the Light Plane* (New York: Commonwealth Books, 1946), pp. 16–17. (2) Holley, pp. 58–59, 62–63, 68, 72–77, 87–88.

6 Holley, pp. 24, 74–76, 86–87.

[7] (1) Ten Eyck, pp. 15–23. (2) Devon Francis, *Mr Piper and His Cubs* (Ames: Iowa State University Press, n.d. [c.1973]), pp. 79–88. (3) Ford, "Forty Years," pp. 4–6. (4) Frederic A. Bergerson, *The Army Gets an Air Force: Tactics of Insurgent Bureaucratic Politics* (Baltimore: Johns Hopkins University Press, 1980), pp. 29–31. (5) Tierney, pp. 42–50, 52–54, 59–61.

[8] (1) W. F. Craven and J. L. Cate, eds., Office of Air Force History, *The Army Air Force in World War II,* Vol. 6: *Men and Planes* (Chicago: University of Chicago Press, 1955), pp. 619–20. (2) Alfred Goldberg, "Between World Wars," in Alfred Goldberg, ed., *A History of the United States Air Force, 1907–1957* (Princeton, New Jersey, 1957) p. 40. Hereafter cited as Goldberg, *USAF History*. (3) Holley, pp. 29–34, 62–64, 77, 82, 89, 92. (4) Russell F. Weigley, *History of the United States Army* (New York: MacMillan Co., 1967), p. 414. (5) Captain Jonathan M. House, US Army Command and General Staff College, Combat Studies Institute, Research Survey No. 2, *Toward Combined Arms Warfare: A Survey of 20th-Century Tactics, Doctrine, and Organization* (Fort Leavenworth, Kansas: GPO, August 1984), p. 131.

[9] Kent Roberts Greenfield and Robert R. Palmer, "Origins of the Army Ground Forces: General Headquarters, United States Army, 1940–42," in Kent Roberts Greenfield, Robert R. Palmer, and Bell I. Wiley of the Historical Sections, Army Ground Forces, *United States Army in World War II, The Army Ground Forces: The Organization of Ground Combat Troops* (Washington, DC: Historical Division, Department of the Army, 1947), p. 111.

[10] (1) Holley, pp. 92–93. (2) Ten Eyck, p. 20.

[11] (1) Laurence B. Epstein, "Army Aviation Logistics in Evolution, 1942–1953, Part 1," *USAAD*, June 1979, p. 8. Hereafter cited as Epstein 1. (2) Holley, pp. 50, 67, 74.

[12] Robert A. Lovett, Assistant Secretary of War for Air, Memorandum for the Chief, Army Air Force, "Procurement Service Quantity of Light Commercial Type Airplanes," 14 August 1941.

[13] (1) Futrell, p. 15. (2) Holley, pp. 97–98, 101. (3) Ten Eyck, p. 22. (4) Air Technical Service Command, Historical Office, "Case History of the Liaison Aircraft Program," Wright Field, November 1944, p. 2, and Document File, no. 9.

[14] Kent Roberts Greenfield, Colonel (Infantry, Reserve), Army Ground Forces, Historical Section, "Army Ground Forces and the Air-Ground Battle Team Including Organic Light Aviation, Study No. 35," 1948, pp. 23, 26. Hereafter cited as Greenfield, "Light Aviation."

[15] (1) Holley, pp. 73, 84. (2) Futrell, pp. 15–16.

[16] (1) Ford, "Forty Years," p. 6. (2) Tierney, pp. 43, 50. (3) Lieutenant Colonel Cecil B. Currey, US Army Reserve, *With Wings as Eagles* (Washington, DC: GPO, 1984), pp. 31–32. (4) Ten Eyck, p. 38. (5) Greenfield, "Light Aviation," p. 23.

[17] (1) Greenfield, "Light Aviation," p. 23. (2) Holley, p. 82.

[18] Francis, pp. 81–83, 85, 87, 92.

[19] Robert M. Danford, Letter to General Salet, Commandant US Army War College, Carlisle Barracks, Pennsylvania 17013, 28 April 1967.

[20] (1) Brigadier General Harry L. Twaddle, Assistant Chief of Staff, Operations and Training Division, G-3, War Department General Staff, Memorandum for the Adjutant General, "Air Observation," 8 December 1941. (2) Rex W. Beasley, Lieutenant Colonel, Field Artillery, Executive, Letter to the Assistant Chief of Staff, G-3, War Department, Washington, DC, "Service Test of Air Observation as an Organic Part of Field Artillery Units," 5 December 1941, paragraph 1. (3) R. G. Hersey, Adjutant General, Memorandum to the Chief of Field Artillery, "Air Observation," 10 December 1941. (4) R. G. Hersey, Adjutant General, Memorandum to the Chief of Army Air Forces, "Air Observation," 10 December 1941.

[21] William W. Ford, Lieutenant Colonel, Field Artillery, Director of Air Training, Memorandum to the Commandant, Field Artillery School, Fort Sill, Oklahoma, "Air Training," 30 April 1942.

[22] Ford, "Forty Years," pp. 9–10.

[23] (1) Brigadier General I. H. Edwards, Assistant Chief of Staff, Memorandum for the Commanding General, Army Air Forces, "Organic Air Observation for Field Artillery," 6 June 1942. (2) Brigadier General I. H. Edwards, Assistant Chief of Staff, Memorandum for the Commanding General, Army Ground Forces, "Organic Air Observation for Field Artillery," 6 June 1942.

[24] (1) Holley, p. 79. (2) Futrell, pp. 15, 17–18.

[25] (1) Futrell, pp. 24–25. (2) Greenfield, "Light Aviation," pp. 61–62, 66. (3) Craven and Cate, Vol. 6, p. 618. (4) Ten Eyck, pp. 63–64.

[26] War Department, Training Circular No. 45, "Employment of Army Air Forces Liaison Squadrons," Washington 25, DC, 11 July 1944.

[27] United States Army Forces in the Far East, Headquarters, Memorandum to the Adjutant General, Washington 25, DC, USAFFE Board Report No. 117, "Army Air Force Liaison Squadrons," 18 February 1945.

[28] Greenfield, "Light Aviation," pp. 98–100.

[29] United States Army Air Forces, Assistant Chief of Staff, Plans, Post-War Division, "Liaison Aviation in the Post-War Military Establishment," 10 September 1945, Tab D, "Combat Reports on Liaison Aviation," Nos. 1–3, 6, 9, 11, 16, 24, 26, 28, 30, 31, 38.

[30] Greenfield, "Light Aviation," pp. 26–28, 58–65, 65–67, 109–110.

[31] Ibid., pp. 26, 27, 58.

[32] (1) Lieutenant General Barney M. Giles, Deputy Commander, Army Air Forces, Memorandum for G-3 and the Chief of Staff, "Organic Assignment of Aircraft Other Than to the Air Forces," 10 October 1944. (2) Brigadier General William W. Welsh, AC/AS, Training, Memorandum to AC/AS, OC&R, "Liaison Type Aircraft," 4 June 1945.

[33] (1) Greenfield, "Light Aviation," pp. 112–13. (2) United States Army Air Forces, "Liaison Aviation in the Post War Air Force," n.d. (3) United States Army Air Forces, Assistant Chief of Staff, Plans, Post-War Division, "Liaison Aviation in the Post-War Military Establishment," 10 September 1945.

[34] Lieutenant General Ira C. Eaker, Deputy Commander, Army Air Forces, Disposition Form, "Liaison Type Aircraft, Comment #2," 3 August 1945.

[35] Greenfield, "Light Aviation," pp. 62–65, 100–108, 113.

[36] Alice C. Cole, Alfred Goldberg, Samuel A. Tucker, Rudolph A. Winnacker, eds., Historical Office, Office of the Secretary of Defense, *The Department of Defense: Documents on Establishment and Organization, 1944–1978* (Washington, DC: GPO, 1978), pp. 1–59. Hereafter cited as Goldberg, *Defense Documents*.

[37] Goldberg, *Defense Documents*, p. 41.

[38] (1) Goldberg, *Defense Documents*, pp. 265, 267–68, 270–85. (2) "Fifteen Months of Unification, First Report of the Secretary of Defense," *AID*, February 1949, pp. 8–25.

[39] War Department, "War Department Equipment Board Report," Washington, DC, 29 May 1946, p. 42.

[40] War Department, Field Manual FM 20-100, *Army Ground Forces Light Aviation* (Washington, DC: GPO, 1947), pp. 122–25.

[41] "Army–Air Force Agreements," *AID*, November 1947, pp. 48–56.

[42] Departments of the Army and the Air Force, Joint Army and Air Force Adjustment Regulation No. 5-10-1, "Combat, Joint Operations, Etc.: Employment of Aircraft for Performance of Certain Missions," Washington 25, DC, 20 May 1949.

[43] Established in 1938, the Civil Aeronautics Authority became the Civil Aeronautics Administration in 1940. Superseded by the Federal Aviation Agency in 1959, it then became the Federal Aviation Administration in 1966.

[44] Epstein 1, p. 10.

[45] (1) "History of Army–Air Force Conflicts Concerning the Function of Organic Aviation," n.d. [c.1974], p. 3. Front cover has hand inscription: "Known as the Williams Report." Hereafter cited as Williams, "Army-AF Conflicts." (2) Department of the Air Force, Memorandum to Commanding General, Air University, Maxwell Field, Alabama, "Liaison Aircraft," 5 January 1948. Hereafter cited as DAF, Memo to CG, AU, "Liaison A/c," 5 Jan. 48. (3) United States Army Military History Institute, Senior Officers Oral History Program, Army Aviation, General Robert R. Williams, Interviewed by Colonel Glenn Smith, Lieutenant Colonel August Cianciolo (Carlisle Barracks, Pennsylvania, 18–19 March 1977), tape 1, pp. 32–33. Hereafter cited as Williams, "Oral History."

[46] US Air Force, Tactical Air Command, compiled by Benjamin Goldman, historian, "History of the Tactical Air Command for the Period 1 January–30 November 1948, Volume 1: Organization & Mission," n.d., p. 98.

[47] (1) Williams, "Army-AF Conflicts," p. 3. (2) DAF, Memo to CG, AU, "Liaison A/c," 5 Jan. 48. (3) United States Air Force, Headquarters, Air University, Project AU-4800, "Aircraft Organic to the Army," Maxwell Field, Alabama, 24 February 1948, pp. 1–3. (4) United States Air Force, Headquarters, Tactical Air Command, "Study of Liaison Aircraft Activities," Langley Air Force Base, Virginia, 19 March 1948, pp. 1–2, 9. (5) Williams, "Oral History," tape 1, pp. 30–31.

SOURCES FOR TABLES

1-1. United States Forces, European Theater, The General Board, "Liaison Aircraft with Ground Force Units," n.d. [c.1945], pp. 8–10.

1-2. See main text.

2

EXPANSION: POSTWAR MODERNIZATION AND THE KOREAN WAR

The immediate post–World War II period was important for the history of Army aviation and air mobility. Air mobility and helicopter usage in the Army began before the outbreak of the Korean war in June 1950. There was one important prerequisite that allowed this to happen. In fiscal year (FY) 1947 the Army began to procure airplanes larger than the two-place liaison-type standard during World War II. The first such airplane was a four-place aircraft with triple the passenger capacity of the two-place airplane. Moreover, it was an interim aircraft, with better airplanes to come, especially after the outbreak of the Korean war. The Army also began to procure two-seat light helicopters. Then in 1949 the Army began to consider transport helicopters. By May 1950 it had decided to form five experimental transport helicopter companies. This meant that Army aviation was moving from the observation role into the mobility role. These developments took place despite the post–World War II cutbacks and demobilization. Friction with the Air Force also did not impede these moves.

The Korean war had several profound effects on Army air mobility development. First, battlefield requirements again acted as an impetus to functional expansion, similar to the World War II process. This was especially evident with the Army using reconnaissance helicopters for medical evacuation. Second, the Marine Corps, more progressive in the field of air mobility, demonstrated its potential in a combat setting. This eliminated the uncertainties surrounding the practical use of transport helicopters. Third, the budgetary expansion of the war gave the Army an opportunity to expand aircraft procurement programs. Fourth, the exigencies of war again thwarted

Air Force opposition to Army aviation expansion. Finally, the Korean war experience utterly convinced top Army officials that air mobility based on transport helicopters would become an indispensable part of the Army. In 1952 the Army high command decided to start a procurement program for 12 transport helicopter battalions each with 67 helicopters divided into three companies. The Korean war was therefore instrumental in stimulating radical air mobility developments.

PRE-KOREA MODERNIZATION

The Army had a considerable degree of freedom to embark on a postwar aircraft modernization and improvement program. The direction to do so came as early as 1946, from the Stilwell Board. Although the Army had to procure these aircraft through the Air Force, this arrangement proved no obstacle. Cost also proved not to be an obstacle, because it involved only a small percentage of the total budget. The birth of Army air mobility might have been the result of Marine Corps activities in this field. Imitation apart, the Army had a requirement to replace parachute delivery of airborne elements. As a result, the Army began to invest research and development funds into heavy lift helicopters and convertiplanes. (Convertiplane was a general term for aircraft with the forward flight characteristics of fixed wing airplanes and with the vertical take-off and landing characteristics of helicopters. Therefore theoretically it had the advantages of both and disadvantages of neither.)

Beyond Two-Place Liaison Airplanes

Two-place fixed wing liaison airplanes remained the mainstay of organic Army aviation during the postwar period. After World War II the Army inherited surplus L-5 airplanes, bought a small number of L-13 airplanes, cancelled the L-14 program, and tested a handful of L-15 airplanes.[1] The Army finally settled on the L-16 Aeronca aircraft in 1948 as a stop-gap measure. This aircraft proved even cheaper than the L-4, costing $2,300 each, with over 700 entering service.[2] Thereafter the Army began a competition to replace all the older two-place liaison airplanes and the L-16. The Army selected the winner of this competition, the L-19 Bird Dog, just before Korea in April 1950. This airplane was so successful that the Army bought thousands and production continued until the sixties.

More important was the L-17 Navion, a four-place airplane with retractable landing gear and steering front wheels. Procured initially in FY 1947 as an interim utility airplane, the Army bought over 200, each costing $10,000.[3] This airplane represented the first step beyond the two-place liaison airplane. Moreover, it was an important test case for the Army. Although the L-17 was still a light aircraft, it was qualitatively much improved. (This was true, even compared with the three-seat L-13, which could take two litter patients by removing the two rear seats.) The increase in capability generated some alarm within the Air Force.[4] However, the Army had no difficulty in procuring the L-17, because this aircraft was an off-the-shelf, unmodified, commercial model. The guidance for this increase in aircraft capability came mainly from the 1946 Stilwell Board. It recommended fixed wing liaison airplanes with more than two seats, as well as liaison-type helicopters of up to three seats.[5]

Helicopters Enter Service

The procurement of helicopters was another instrumental step in Army aviation development. The Army's strong interest in the helicopter dated back to World War II, when the Army Air Forces procured several hundred experimental R-4, R-5, and R-6 helicopters. Although the AAF used some of them in the China-Burma-India theater, the Army Ground Forces never operated any. In 1945 a group of Army pilots visited the Bell Helicopter Company and observed a demonstration of the light two-seat YH-13 model. The responsiveness of the YH-13 impressed them enough that they pressured their superiors to procure adequate numbers for experimentation. The 82nd Airborne Division received seven YH-13 helicopters in December 1946 for evaluation and test. The initial tests revealed the great versatility of this helicopter, and Army officers urged additional procurement.[6] Because of this favorable response, on 5 August 1947 General Devers (Commanding General, Army Ground Forces) requested General Eisenhower (Chief of Staff) to authorize the procurement of 150 light helicopters. He also requested buying 50 of the same helicopters for the Air Force, as it was responsible for primary flight training for all Army aviators. Finally he recommended the development of cargo helicopters as well.[7]

Although 150 light helicopters seems a modest number, their high unit cost of $45,000 made a large procurement beyond the finances of the Army aviation budget (see below for details).[8] On 29 August 1947 representatives of the Army Ground Forces and the Army Air Forces agreed to reduce other types of Army aircraft procurement appropriately. The Army Ground Forces was to submit a revised FY 1948 aircraft program based on this conference. The Acting Chief of Staff, Lieutenant General Lawton Collins, approved the request on 8 September 1947.[9] In the end the Army could find money in 1948 for only 65 H-13B helicopters. In 1950 the Army ordered 86 H-13D helicopters, reaching the 150 total originally envisaged, as well as ordering another 50 two-place H-23 helicopters. These helicopters did not provoke controversy with the Air Force, as they had only two seats and counted as observation aircraft. In addition, the H-13 was a CAA-approved aircraft.

Even though the Army had a limited number of H-13 helicopters, it nevertheless began to distribute them to field units by 1949. The school troops of the Infantry, Armored, and Artillery Schools, and the Army Field Forces (the major command in the United States) also received the H-13. The Army used them for artillery fire adjustment, observation and reconnaissance, courier service, traffic and march column control, front line patrol, wire laying, and even for limited short-range supply.[10] In summer 1949 the Army also began to investigate the use of helicopters for rescue and evacuation, a role the Stilwell Board report recommended in 1946.[11] The H-13 had an inherent capability as a helicopter ambulance. It was capable of carrying a pilot, medical attendant, and two external litter patients. On observation missions it normally had a crew of two.

By comparison, the Army did not place any procurement orders for transport helicopters before the Korean war started, because none were available.[12] Neither the Stilwell Board nor the 1950 follow-on Hodge Board listed larger transport-type helicopters as a requirement. Nevertheless, the Army, following General Devers' recommendation, asked the Air Force to develop a helicopter with a $2^1/_4$–ton payload. The Air Force fully recognized the importance of developing transport helicopters and convertiplanes, especially to replace the C-46 and C-47 medium cargo airplanes.[13] The Air Force Chief of Staff, General Hoyt Vandenberg,[14] and General Robert Lee of

Table 2-1
Army Aviation Budget, FY 1948–1950,
Obligations Incurred, Ordnance Department

Fiscal Year	Procurement of Liaison Aircraft	Maintenance of Liaison Aircraft	Total Direct Obligations	Percentage of Obligations
1948	$4,263,716	$569,372	$220,056,991	2.2
1949	939,776	2,004,637	586,932,859	0.5
1950	7,320,964	2,368,017	627,202,695	1.5

the TAC both held this belief.[15] However, the Air Force had to develop them under a restricted schedule because of the postwar resource constraints. Moreover, fixed wing assault transports and large troop carriers had a higher priority.[16]

Resources for Aviation Modernization

At the peak of World War II the Army consisted of 69 infantry, 20 armored, and 5 airborne divisions. There are various estimates that the Army had 1,600, 1,700, or even 3,000 airplanes in 1945.[17] Rapid demobilization in the immediate postwar period reduced the number of divisions drastically to only a handful. Defense expenditure was only a fraction of the war years. However, the Army still managed to find the funds for new and improved (and therefore more expensive) aircraft such as the L-17 and the H-13. Moreover, the Army maintained its aircraft inventory at a high level, considering the decline in the number of divisions. On 16 December 1947 the Army had a worldwide total of 681 aircraft including helicopters.[18] This number increased to over 700 by June 1950 (see Table 2-2).

There are several explanations for this. First, there were more aircraft assigned to ground units. In August 1945 the War Department approved the extension of Army aviation to other ground units. This would increase the 10 aircraft in an infantry division to 16, and the 8 aircraft in an armored division to 17 (see Chapter 1). A divisional reorganization in 1947 implemented this increase, although sources are vague as to the exact number.[19] Each infantry division may have got 16 liaison airplanes, while each armored division may have eventually got 21 liaison airplanes.[20] The new table of organization and equipment may have included the H-13B helicopter.[21]

Second, the Army's aircraft procurement program was still affordable. It represented a small percentage of the overall budget, even though it included more expensive aircraft (see Table 2-1). The FY 1948 to FY 1950 budgets alone were more than enough to buy 65 H-13B helicopters at $45,000 each, 700 L-16 airplanes at $2,300 each, and 246 L-17 airplanes at $10,000 each. (No breakdown exists for the FY 1946 and FY 1947 figures, which came under the AAF procurement budget.)

Instead of the traditional budgetary cycle of "feast and famine" experienced during war and peace respectively, the armed forces were experiencing the highest peacetime readiness in history.[22] The Army, not suffering the worst "famine" after World War II, therefore had more room to undergo aviation modernization and expansion. However, it was a financially stringent period for the rest of the Army. The postwar occupation burden took up a great part of both personnel and funds. This forced the Army to get along with surplus World War II weapons and equipment, except for missile development.[23]

Perhaps the Army was willing to invest in aviation capabilities while neglecting other areas because the Army high command realized aviation was an important component of land combat. Hence it was trying to obtain an increased capability in this area.[24] In 1949 Army Chief of Staff General Omar Bradley stressed the need for a tactically and strategically mobile Army.[25] In November 1949, Lieutenant General Matthew Ridgway, Deputy Chief of Staff for Administration, said that "the Army plans to place increasing emphasis upon airborne, air-transportability and air-ground support techniques."[26] The move to establish an initial air mobile capability perhaps indicated this desire.

The Dawn of Air Mobility

The Marine Corps was the first to seize the potential offered by the helicopter. During the immediate post–World War II period it formulated the general theory of air mobility for amphibious warfare. This was necessary because nuclear weapons were endangering its traditional role by threatening an amphibious fleet massed before an assault landing. The 1946 Bikini Atoll nuclear test affirmed the need for a new approach to amphibious operations. Thereafter, the Marine Corps set up a board to examine other options. It recommended using helicopters to provide dispersion during the approach to the beachhead while enabling a force to concentrate at the moment of assault. The Commandant of the Marine Corps accepted the findings of this board, which became a future planning guide. The Marine Corps then began to work out the details and began validating the concept during various exercises. It also formed the first helicopter transport unit during this period. Only the equipment was lagging behind, not only in terms of helicopters but also helicopter-carrying assault ships. The Marine Corps strove to obtain the required matériel from this point on.[27]

By contrast the Army was relatively slow to establish an air mobile capability. However, this began to change from 1949, when the mainstream Army and the airborne element began to take an interest. Late in 1949 the Office of the Chief of Army Field Forces

(OCAFF) conducted studies that indicated the need for various types of helicopters. They were to provide short-haul air transport to corps, divisions, and smaller tactical units of the ground forces. This proposal received strong support in a study prepared by the Department of the Army, G-3, in May 1950. A recommendation to provide funds in the FY 1952 budget for organizing and equipping five transport helicopter companies followed. Placed with divisions in the United States, they would develop doctrine for their employment. The Army was to implement a full-scale program after the experimental five-company program. In summer 1950 the Transportation Corps received a role, largely of a staff nature, in the development of this program. Its functions involved planning and coordinating the activation and equipping of cargo helicopter units with the Army Field Forces and the Army general staff. It also involved formulating doctrine for the employment of cargo helicopter units.[28]

The airborne community's interest in transport helicopters surfaced during the April 1950 Army Airborne Panel chaired by Major General William Miley. The basis of its interest stemmed from the unsatisfactory nature of the parachute. The panel reported:

> Troop type parachutes have inherent limitations which seriously restrict their adequacy for airborne use. Vulnerability during parachute descent, together with the difficulty of assembling scattered troops upon landing, demonstrate that this method for landing troops should be superseded by a more effective one. By the end of fiscal year 1951, troop type parachutes will have been developed to their ultimate effectiveness within the restrictions of their inherent limitations. For this reason, no further development effort is justified after that time.[29]

The panel concluded that the mobility and flexibility offered by the convertiplane and the heavy lift helicopter would increase the Army's fighting potential. It therefore recommended that both should receive energetic Army support in their development. The panel saw the convertiplane as a future replacement for transport airplanes.[30] In justifying these recommendations, it said:

> Heavy lift helicopters now under development are expected to improve the Army's flexibility and mobility by providing a short-range vehicle for

> crossing equipment and troops over rivers, defiles, impassable terrain, or other obstacles. Such helicopters warrant development to their full effectiveness. . . . The successful completion of this [convertiplane] project is the most important Army requirement in the field of air transportation, and is absolutely essential to the full realization of airborne potential. . . . The convertiplane is of overriding importance because airborne operations with the conventional type of air transportation . . . are inherently hampered by the necessity of airfields. Airborne operations which are tied to existing airfields or encumbered by the inescapable delays of construction of new airfields will never realize the complete flexibility and mobility which are the true pinions of the airborne concept. A convertiplane transport which will rise and descend vertically, and fly with acceptable forward speed, will free airborne operations from their present transport encumbrances and permit them to become the Army's mode of operation in the future.[31]

The Department of the Army, G-3, concurred with the panel's research and development recommendation, as long as it was in line with the Army's overall priorities. On 3 August 1950 the G-3 staff made a presentation of the panel report and its own comments to the Vice Chief of Staff, General Haislip. On 11 August 1950 General Lawton Collins, who became Army Chief of Staff on 16 August 1949, also received a presentation. Both men approved. These presentations resulted in extra funds for Army convertiplane research and development.[32] Even the onset of the Korean war did not change the basic recommendations and the implementation of the outline program.[33]

THE EFFECTS OF THE KOREAN WAR

Army aviation experienced another significant expansion during the Korean war. This involved not only the aviation budget, aircraft procurement, and the aviation force structure. It also involved the procurement of more capable aircraft and an expansion of aviation functions at Air Force expense. The most salient indicator of this expansion was the 1952 plan to acquire a force of 12 transport

Table 2-2
Army Aircraft Inventory, FY 1950–1952

Fiscal Year	Fixed Wing Aircraft 2-Place	3-Place+	Rotary Wing Aircraft Utility	Cargo	Total
30 June 1950	525	143	57	0	725
30 June 1951	843	165	86	0	1,094
30 June 1952	1,461	271	320	1	2,053
30 December 1952	1,534	320	647	72	2,573
Undelivered 31 December 1952	629	188	556	110	1,483

helicopter battalions. There was a combination of factors that led to this expansion, namely the need to satisfy battlefield exigencies and the availability of a large budget. The Marine Corps' combat experience with helicopters gave a further impetus for Army helicopter development. Moreover, the Army was able to overcome Air Force opposition.

Battlefield Requirements

During the Korean war there was a further increase in the number of divisional and nondivisional aircraft (circa 1952). Each infantry or airborne division now had 26 aircraft, while each armored division had 28. Apart from the 13 liaison-type observation airplanes (15 for armored divisions), there were also 3 utility airplanes of the L-17 type and 10 reconnaissance helicopters. The Army still assigned these aircraft organically within each division, without forming any aviation units. Separate, nondivisional units and higher headquarters also had additional aircraft assigned.[34] This, and the increase in the Army force structure, led to a rapid growth in the Army aircraft inventory (see Table 2-2).

Army fixed wing airplanes in Korea still performed conventional missions such as reconnaissance (observation), courier, and the transportation of commanders and staff officers. They also located targets for air strikes as well as leading fighter-bombers into the attack. More important, they also performed minor movement of equipment, supplies, and troops. Two important fixed wing airplanes with relatively significant cargo capacity, the L-20 Beaver and the L-23 Seminole, entered service in 1951 and 1952. These two acquisitions represented another important advance for Army aviation. The seven-place (six-passenger) L-20 (which could hold three litter patients plus a medical attendant) was a replacement for the interim L-17 airplane. The twin-engine, six-place (four-passenger) L-23 became a standard transport for commanders and staff officers. It could also convert into a cargo or litter patient carrier. Both airplanes could drop small parachute cargoes.[35]

The acquisition of larger payload-carrying aircraft was important because this capability was a prerequisite for developing air mobility. The relative increase in cargo-carrying aircraft came about because of the terrain of Korea, which hindered ground transport. General Collins told Congress in 1953: "Korea has emphasized once again the vital necessity for air movement in the battle zone. At times, almost impassable terrain and inadequate roads and railroads have made aerial transportation the only means of supporting our frontline troops."[36]

While the Army had a growing number of fixed wing airplanes capable of carrying cargo and passengers, helicopters became increasingly predominant. Some terrain in Korea demanded a vertical take-off and landing capability. The only other effective alternative was to travel on foot or by mule. Even with a short rough-field capability, Army airplanes could not land or take off from some sites. This factor profoundly influenced the direction of air mobility development. It brought home the differences between fixed wing and rotary wing aircraft and the fact that the Army may not operate on gentle or rolling terrain with good transport infrastructure. General Collins said in 1952:

> In Korea, we have seen time and again the inestimable value of resupply by airdrop in fast moving tactical situations. The rugged terrain and the inadequate roads and railroads there have necessitated air supply as the best means in many instances of supporting our troops.
> Our helicopters have become particularly important in such work [as well as many other missions]. . . .
> The improvements in our helicopters have progressed to the extent that we now find them a necessary means of supplementing our jeeps and trucks. Helicopters are now an integral part of our organic transportation.[37]

As a result of the terrain factor, the number of helicopters expanded rapidly in the Army inventory (see Table 2-2). This was especially true of the two-place reconnaissance helicopter, which also acted as a four-place utility helicopter. Nowhere was the Army's shift to helicopters more apparent than in the urgent call for air ambulances. The rapid identification of this need and the subsequent formalization of this role were almost entirely a result of expediency. Ground elements requested helicopter medical evacuation from Air Force air rescue detachments as early as June 1950. These requests grew in volume because of the difficulties of evacuating patients by ground vehicles. The Eighth Army developed a keen interest in this operation and conducted a joint test with the Air Force on 3 August 1950. Soon afterward the Surgeon General of the Army, then on a theater tour, requested the dispatch of a substantial number of helicopters for evacuation missions. The first such Army detachment arrived in November 1950 equipped with the H-13. After modifying the helicopters for carrying external litters, the detachment became operational on 1 January 1951 and performed the first evacuation on 3 January.[38] Helicopter ambulances not only had the advantage of overcoming ground transport problems, they also speeded up the delivery of the wounded to hospitals. General Collins explained:

> Army helicopter ambulances have saved the lives of more than 8,000 critically wounded men. In the early days of Korea, before helicopters were available, it sometimes took 12 to 14 hours for teams of litter bearers to descend precarious mountain trails to rear area medical stations. With only first-aid treatment the long trip often proved fatal to the wounded. But with Army helicopters now in the theater, wounded have been picked up from seemingly inaccessible spots and carried to hospitals in as little as 5 minutes.[39]

This speed of recovering casualties improved their chances of survival. From 1 July 1952 to 30 June 1953, 977 of every thousand wounded who reached battalion aid stations survived. This was a substantial increase compared to a World War II figure of 956, and 919 for World War I (excluding gas figures).[40] From September 1951 to January 1953, three provisional Army helicopter detachments operating a total of 12 helicopters had evacuated nearly 11,000 casualties from the battlefield. For the war as a whole, helicopter detachments (including Air Force detachments) assigned to the Army medical service evacuated 21,212 wounded personnel, using only about 20 helicopters. This did not include evacuations made by divisional reconnaissance helicopters. Because of other urgent duties, they did not keep records of the number of patients flown.[41] Once the helicopter proved itself in this role, the Army embraced its inclusion in the future force structure. It also overshadowed the use of fixed wing aircraft for evacuation, which so far had only been an aircraft specification and not a role routinely performed on such a large scale. Moreover, the Korean war experience determined the basic procedure for future emergency and routine casualty evacuation.[42] The use of helicopter ambulances was so convincing that in 1953 General Collins proclaimed: "This battlefield evacuation has become so clearly an everyday necessity that we have recently activated helicopter ambulance organizations as an integral part of our Army Medical Service Corps, and we are now planning 12 of these units to be attached to our divisions and corps as needed."[43]

Transport Helicopter Expansion

While the Army decided to form five experimental companies before Korea, it did not obtain its first transport helicopter until 1952. Moreover, the Army admitted using the experiences of "other services" to publish the first table of organization and equipment for a transport helicopter company on 24 October 1950. Furthermore, the Army's concept of transport helicopter usage had not fully developed at this point.[44] The first doctrinal publication on the use of these transport helicopter units did not appear until June 1952, written by the Transportation Corps.[45] It was up to the Marine Corps to demonstrate the helicopter's potential in land warfare. It had the first transport helicopter squadron in the Korean theater (August 1951). It performed several impressive operations before the first Army helicopter transport unit arrived in December 1952. Several operations received high media profile and military attention and praise, notably operations SUMMIT, BUMBLEBEE, and HAYLIFT II.[46]

By comparison the Army transport helicopter units accumulated much less operational experience. Apart from the Marine Corps operations the Army also had to extrapolate from its own medical evacuation and smaller utility helicopter operations.[47] Even experience gained in World War II was helping to determine the need for transport helicopters.[48]

The Korean war accelerated the process of assimilating transport helicopters into the Army by defining the role of transport helicopters and the force structure required. During this period various Army agencies proposed their own study of a full-scale program (see Table 2-3). The study conducted by the Materiel Requirements Review Panel in July 1952 was the most influential. The Office of the Army

Table 2-3
Summary of Army Transport Helicopter Studies, 1950–1952

Date	Agency	Core Recommendations
May 1950	ACS, G-3	5 experimental transport helicopter companies.
8 Oct. 1950	ACS, G-3	4 transport battalions + 19 companies; or 1 battalion per corps, 1 company per division, + 2 companies in Alaska.
10 Aug. 1951	OCT	19 light or 8 medium or 4 large cargo battalions per field army.
13 Nov. 1951	CG, FECOM	4 battalions for Korea (20 H-23, 260 H-21), 10 battalions per field army.
7 July 1951	Joint Airborne Troop Board	3 general cargo, 1 heavy lift battalions per field army.
11 July 1952	OCAFF	6 light, 3 medium, 1 heavy battalions per field army.
18 July 1952	OCT	50 helicopters per overseas division slice, 3–500 in US; or 1,500 cargo helicopters.
31 July 1952	Materiel Requirements Review Panel	15 cargo battalions (3 companies of 21 transport helicopters each).
8 Jan. 1953	OCT	12 battalions, or 41 helicopters per division slice (28 medium & 13 light).

Chief of Staff initiated this study on 30 June 1952. The panel called for the formation of 15 transport helicopter battalions each with 3 companies of 21 transport helicopters. For planning purposes each field army would have 3 light, 3 medium and 1 heavy transport battalions. The total cost would amount to $1 billion over five years. The panel also made many other far-reaching recommendations. For example it urged the Army to assume all responsibility for depot supply and maintenance as well as training. It also called for the Army to initiate a five-year helicopter research and development program, investing $15 million per year.[49]

The proposed increase was obviously massive. General Collins received a briefing on 21 August and gave approval for only 12 battalions, perhaps because of the costs involved. He issued a directive to have these established by 1956 or as soon thereafter as possible. However, without an expansion of production facilities the Army could not accomplish the program until 1959. The Army also needed to divert huge funds from other programs to finance the total program cost, now estimated at $688.8 million. Moreover, some aspects of the proposed program remained vague and proved difficult to incorporate into yearly budget estimates. The Secretary of the Army ordered the Office of the Chief of Transportation to conduct another staff study to remedy this situation. Published on 8 January

Table 2-4
Cost Comparison of Various Helicopters, May 1952

Category	Model	With Spare Parts	Without Spare Parts
Light Cargo ($1\frac{1}{2}$–ton)	H-21	$530,343	$307,202
Utility	H-19	274,060	157,080
	H-25	269,000	178,000
Reconnaissance	H-23	64,281	40,583
	H-13	52,136	31,358

1953 and approved by the Vice Chief of Staff, the study provided three main detailed justifications for the 12 battalions calculated on a division slice basis. Each division required 15 medium and 7 light transport helicopters for the logistical mission. For the tactical mission each division required 10 medium and 5 light transport helicopters. Each division also needed 3 medium and 1 light transport helicopters for the movement and handling of special weapons. These helicopters would be able to move 20 percent of a division's daily supplies (plus other logistical needs) and one reinforced infantry battalion in two lifts. This could reduce the number of trucks needed by 14 percent.[50]

This large force structure proposal was justifiable because there was a massive Army expansion due to the Korean war. While its prewar personnel strength hovered around the 600,000 mark, during the war it stood above 1.5 million. The number of active divisions also rose commensurately from 10 under-strength divisions to 20.[51] While the sum was astronomical, the proposed increase in transport helicopters was not an unbearable drain on Army procurement funds. The war meant that there was a dramatic increase in the Army budget. This could support a large inventory growth and more expensive aircraft (see Tables 2-4 and 2-5).[52] This was sustainable only if funding remained at a high level.

Besides the 12 transport helicopter battalions, Army plans called for the acquisition of a complete range of helicopters. The first was the two-place reconnaissance helicopter used for reconnaissance, training, observation, courier, and liaison roles. A second type would be a utility helicopter for the medical evacuation role, as an instrument trainer, and for any general utility missions beyond the first type's capabilities. Besides carrying one pilot, it would have an 800-pound cargo capability, or internal accommodation for a minimum of two standard Medical Corps litters and one medical attendant. The third type was a family of cargo helicopters with $1^1/_2$–ton, 3-ton, and 5-ton payloads. They were to carry their full cargo load externally and full passenger capacity internally, while capable of

Table 2-5
Army Aviation Budget, FY 1951–1953,
Obligations Incurred, Procurement and Production, Army

Fiscal Year	Army Aircraft	Total Direct Obligations	Percentage of Obligations
1951	$118,120,364	$7,042,758,532	1.7
1952	90,998,226	8,970,571,591	1.0
1953	37,000,000	2,420,789,799	1.5

quick conversion to carry a maximum of standard litters. The fourth category was the heavy lift helicopter with a 16,000-pound cargo capability. This was for moving heavy items such as light tanks, special vehicles, artillery weapons, and general cargo and equipment.[53]

The Pace-Finletter Agreements

Despite the earlier agreements between the Army and the Air Force, areas of contention surfaced again quickly as a result of the Korean war and the expanding role of Army aviation. There were two main areas of contention. One concerned the size and weight of aircraft the Army wanted to procure. They were beyond the 2,500-pound fixed wing and 4,000-pound rotary wing limits set by JAAFAR 5-10-1. The other involved the Army's desire to expand its aviation functions.

The Air Force opposed the Army request for aircraft exceeding weight limits by deliberately delaying such procurement. The aircraft involved included the L-20, L-23, H-19, and H-21.[54] The Air Force also began to set up its own helicopter assault squadrons. TAC made the request during August and September 1950. The Department of the Air Force approved in November 1950.[55] This was not only to compete with or head off the Army's calls for operating its own transport helicopters. It also meant the Air Force was buying all the limited production quantities of transport helicopters, leaving none for the Army. These disputes reached the departmental level. On 24 January 1951 the Army Secretary, Frank Pace, Jr, proposed a replacement of the JAAFAR 5-10-1 to his Air Force counterpart, Thomas Finletter. The Air Force refused.[56] A second high-level discussion occurred on 16 February 1951 with General Collins, Army Vice Chief of Staff General John Hull, Air Force Under Secretary John McCone, and General Hoyt Vandenberg. Vandenberg, the Air Force Chief of Staff, agreed only to allocate the requested helicopters on an experimental basis.[57]

This still left the issue of L-20 procurement unresolved.[58] On 1 August 1951 a McCone memorandum indicated willingness to waive weight restriction for L-20 procurement while reaffirming JAAFAR 5-10-1.[59] Then on 2 October 1951 the two services made a new agreement, the so-called 1951 Pace-Finletter agreement. This new interservice agreement realigned the dividing line between the two services in favor of the Army. It not only abolished weight restrictions on Army aircraft, it also expanded the role of Army aviation (see Table 2-6).[60]

There are two possible explanations for the sudden Air Force willingness to compromise. One was its inability to provide what the Army wanted. The Air Force liaison squadrons, still nominally its responsibility, had all but disappeared. There is no indication that the Air Force was ready to operate small staff transports, such as the L-23, for the Army. Note that the Air Force gave up the liaison functions in the agreement. Another but less likely explanation was the weak Air Force hand in providing tactical air support. This was making the Army call for a larger voice in tactical air support issues.[61] Perhaps because of this difficulty, the Air Force gave way to Army demands. The alternative might have been to jeopardize its tactical air power functions altogether in the long term, or at least lose operational control over its units.[62] Congress and the executive could conceivably give the Army the tactical air support role if dissatisfaction with Air Force negligence became wide and deep enough.[63]

Despite the 1951 Pace-Finletter agreement, battlefield exigencies soon made the agreement virtually obsolete. Areas of contention included more fixed wing aircraft and helicopter procurement requests, the allocation and role of Army and Air Force helicopter transport, and aeromedical evacuation. Exchanges of correspondence between top Army and Air Force officials ensued.[64] The lack of Air Force flexibility induced Pace into a staff reevaluation of Army aviation, which the Materiel Requirements Review Panel undertook.[65] The Air Force responded negatively to the panel's report. Civilian defense officials had to arbitrate in this dispute. On 13 October 1952 the two service secretaries and chiefs of staff met along with a deputy secretary of defense. They could not reach any agreement. They therefore referred the matter to Secretary of Defense Robert Lovett.[66] Under his auspices, the two services signed a second memorandum on 4 November 1952.[67] While the agreement reintroduced a weight limitation of 5,000 pounds for fixed wing

Table 2-6
The 1951 and 1952 Pace-Finletter Agreements

Army Aviation Functions:
• Aerial observation to amplify and supplement other Army methods of observation for the purpose of locating, verifying and evaluating targets, adjusting fire, terrain study, or obtaining information on enemy forces not otherwise obtained by air reconnaissance agencies of the other Services; this includes limited aerial photography incident to these purposes.
• Control of Army forces.
• Command, liaison and courier missions pertinent to the combat area and training therefor.
• Aerial wire laying within the combat area.
• (1951) Transportation of Army supplies, equipment and small units within the combat zone...though it is recognized that the Air Force is assigned the primary function of supplying the necessary airlift to the Army.
• (1952) Transportation of Army supplies, equipment, personnel and small units within the combat zone.
• (1952) Aeromedical evacuation within the combat zone, to include battlefield pickup of casualties, their air transport to initial point of treatment and any subsequent move to hospital facilities within the combat zone.
• (1952) Artillery and topographic survey.
Air Force Logistical Air Support (1952):
• Airlift of Army supplies, equipment, personnel and units from exterior points to points within the combat zone.
• Airlift for the evacuation of personnel and materiel from the combat zone.
• Airlift for the air movement of troops, supplies and equipment in the assault and subsequent phases of airborne operations.
• Aeromedical evacuation for casualties from the initial point of treatment or point of subsequent hospitalization within the combat zone to points outside of the combat zone; and in airborne operations, the evacuation of all casualties from the objective area until such time as ground link-up is attained.

aircraft, it was favorable to the Army as the 1951 version had been (see Table 2-6).[68] Afterward, the Army Chief of Staff, General Collins, publicly assured that the Army did not intend to create another "air force" or duplicate the functions of the Air Force. The transport helicopter companies were to replace truck companies.[69] Moreover, the new agreement resolved the problem with aeromedical evacuation.[70]

Appreciation by Top Officials

Throughout the Eighth Army in Korea, officers' testimony recognized and appreciated the value of transport helicopters.[71] Korea had opened military commanders' eyes to the tactical potential of helicopters. They now awaited the development of new prototypes that would give them unprecedented capabilities for future combat operations. They were also hopeful of convertiplane development.[72] Moreover, public pronouncements by top Army civilian and military officials indicated how much they appreciated the role of Army aviation and transport helicopters in land warfare. The Assistant Chief of Transportation for Army Aviation, Colonel Robert Neely, said in 1954: "Not only the aviation enthusiast but the command echelon of the Army immediately recognized the potential of the larger helicopter. In fact the enthusiasm for the helicopter usually became higher with the rank of the officer."[73] For example, General Collins told Congress as early as 1952:

> We are making great strides in the application of air mobility to Army tactics on the battlefield. Our Army can move faster and farther and hit an enemy with greater surprise than ever before because of our advance in air mobility. This is a field which is rapidly developing and one in which I feel we must place ever increasing emphasis. It is my firm conviction that in any future war airborne operations will play a greater part than ever before, and I am pleased to tell you that air transport is greatly multiplying the Army's potential effectiveness.[74]

Two Eighth Army commanders who later became Army Chiefs of Staff also recognized the value of helicopters. General Matthew Ridgway, then the Commanding General of Far East Command (FECOM), said in the Budget and Plans for FECOM dated 13 November 1951:

> Operations in Korea have conclusively demonstrated the vital necessity for this vehicle in the conduct of modern warfare, especially in mobile situations.[75]

On the same day Ridgway proposed a transport helicopter plan involving four battalions for Korea, and ten battalions per field Army (see Table 2-2).[76] Before 10 November 1951 he also asked for the immediate dispatch of 122 transport helicopters to Korea.[77] In October 1953 General Maxwell Taylor commented on two experimental logistical helicopter operations carried out in Korea, SKYHOOK and SKYHOOK 1:

> These operations have left little doubt as to the capability of the cargo helicopter to fly units and

> supplies to points beyond the means of other methods of transportation. Subsequent combat operations have tended to substantiate this belief. The cargo helicopter, employed in mass, can extend the tactical mobility of the Army far beyond its normal capability. I hope that the United States Army will make ample provisions for the full exploitation of the helicopter in the future.[78]

Colonel William Bunker, Air Transport Service Division, Office of the Chief of Transportation, predicted that helicopters in future would replace trucks to a significant extent. He reckoned that eventually heavy lift helicopters would move 80 percent of the Army's troops and equipment and allow a mobility and dispersion never known before.[79] Brigadier General Paul Yount, the Acting Chief of Transportation, said in January 1954:

> It now seems clear that the helicopter is destined to play an increasingly important role in the tactics and logistics of the ground Army in the future. . . . While the helicopter of today still has a long way to go to become an ideal vehicle, its already tremendous margin of superiority over traditional Army vehicles is an ample promise of the future.[80]

Top civilian officials such as the Secretary and Under Secretary of the Army also appreciated the critical role of helicopters in the future.[81] Secretary of the Army Frank Pace, Jr, told the American Helicopter Society on 15 May 1952:

> So valuable has the helicopter become for transporting men, supplies and equipment in Korea, that we have established transportation helicopter companies as standard units in our Transportation Corps. We have also added helicopters to the standard transportation of many other Army units.
>
> There is no doubt in my mind that when the lessons of Korea have been fully assessed, the dependence of ground units on helicopters for observation, communication, control, and fire direction—as well as for transport—will emerge as one of the most important. And, from the Korean battleground,

> commanders continue their urgent requests to "send more Army aircraft—especially helicopters!"[82]

Under Secretary of the Army Earl Johnson's interest in the helicopter was equally strong.[83] Addressing the Dallas Council on World Affairs on 20 November 1953, he highlighted the emphasis that the Army was placing on what he called "this new instrument of airpower." Furthermore, he said:

> It is obvious from the past studies which we have conducted in the Army at various staff levels, in the field and during operations in Korea, that the helicopter is a tool admirably adapted to meet many Army requirements. These studies, detailed as they are, can do no more than point the way. I am sure that many uses will be found for the helicopter which we do not now even visualize. Despite this fact, we already know that it is of such importance as to warrant including it as an organic vehicle in the Army's operation. Consequently, we have a deep and continuing interest in the development of helicopters.[84]

In short, the Army high command fully appreciated the benefits of transport helicopters in land warfare. It did not feel that the transport helicopter plan involved any excessive risk or uncertainty. However, the Army high command did not have full knowledge of their operating costs, specifically with the high cost of parts and maintenance. Moreover, the decision makers were overoptimistic about the state of helicopter technology and production capabilities. These problems came back to haunt them in the post-Korea period.

NOTES

1 "Army Aviation Equipment," *Army Aviation* 33, No. 12 (31 December 1984), pp. 29, 32–33, 36.

2 House of Representatives, Hearings Before the Subcommittee of the Committee on Appropriations, *Department of the Army Appropriations for Fiscal Year 1951* (Washington, DC, 1950), 81st Congress, 2nd Session, p. 806. Hereafter all congressional appropriations hearings cited as Senate or House of Representatives, *Army (Air Force or DoD) Appropriations FY (year) (part)*.

[3] Official Publication of the Aircraft Industries Association of America, Inc., *The Aircraft Yearbook for 1949*, 31st annual ed. (Washington, DC, 1949), p. 240.

[4] (1) Williams, "Oral History," tape 1, pp. 32–33. (2) Major General Robert M. Lee, CG, TAC, Presentation to the Fairchild Board, "Troop Carrier Aircraft Requirements," 17 May 1949, p. 16.

[5] War Department, "Equipment Board Report," p. 42.

[6] (1) Colonel John R. Galvin, *Air Assault: The Development of Airmobile Warfare* (New York: Hawthorn Books, 1969), pp. 255–56. (2) Tierney, pp. 205–6, 248, 253–55. (3) Lieutenant General James M. Gavin, *War and Peace in the Space Age* (London: Hutchinson, 1959), p. 112.

[7] General Jacob L. Devers, Commanding, Office of the Commanding General, Headquarters, Army Ground Forces, Fort Monroe, Virginia, Memorandum to the Chief of Staff, United States Army, Washington 25, DC, "Purchase of Helicopters for Assignment to AGF," 5 August 1947.

[8] House of Representatives, *Army Appropriations FY 51*, p. 806.

[9] Lieutenant General J. Lawton Collins, Acting Chief of Staff, Memorandum to the Commanding General, Army Ground Forces, Fort Monroe, Virginia, "Purchase of Helicopters for Assignment to Army Ground Forces," 8 September 1947.

[10] "The All-Service Helicopter," *AID*, June 1949, pp. 13–18.

[11] Tierney, pp. 206–7.

[12] Office, Chief, Army Field Forces, Army Equipment Board, "Report of the Army Equipment Board," Fort Monroe, Virginia, 8 March 1950, pp. 88–89. Hereafter cited as OCAFF, "Report of the Army Equipment Board."

[13] (1) United States Air Force, Air Material Command, Intelligence, T-2, Historical Division, "The AAF Helicopter Program," Wright Field, October 1946, pp. 50–51, 59–61, 93. (2) United States Air Force, Air Material Command, Executive Secretariat, Historical Office "History of the Air Material Command, 1 January–30 June 1950, Volume I – Text," Wright-Patterson AFB, October 1950, pp. 145, 153–54. (3) United States Air Force, Tactical Air Command, compiled by Benjamin Goldman, historian, "History of the Tactical Air Command for the Period 1 January–31 December 1949," n.d., p. 108. Hereafter cited as TAC, "History 49." (4) United States Air Force, Tactical Air Command, compiled by Benjamin Goldman, historian, "History of the Tactical Air Command, 1 January–30 June 1950, Volume 1: Narrative & Supporting Documents," n.d., pp. 43–44.

[14] Department of the Air Force, *Report of the Chief of Staff of the Air Force to the Secretary of the Air Force, 30 June 1948* (Washington, DC: GPO, 1948), pp. 78–79.

[15] Major General Robert M. Lee, Speech at the Sixth Annual Forum Meeting, AH Society, Philadelphia, "The Helicopter—Its role in Tactical Air Power," 31 March 1950, pp. 4–5.

[16] (1) Alfred Goldberg, "The Worldwide Air Force," in Goldberg, *USAF History*, pp. 111–12. (2) Maurer, p. 10. (3) Galvin, p. 256. (4) TAC, "History 49," p. 108. (5) Lee, "Troop Carrier A/c Requirements," pp. 12–13.

[17] (1) Epstein 1, p. 9. (2) Lee, "Troop Carrier A/c Requirements," p. 8. (3) Lieutenant Colonel Jack L. Marinelli, G-3 Section, Office, Chief, Army Field Forces, "Army Aviation Surveys the Battle," *AID*, April 1950, p. 49.

[18] DAF, Memo to CG, AU, "Liaison A/c," 5 Jan. 48, Inclusion 1.

[19] (1) House, pp. 146–49. (2) Lieutenant General C. P. Hall, Director of Organization and Training, General Staff, US Army, "Organizing and Training the New Army," *AID*, November 1947, pp. 3–6. (3) Weigley, pp. 502–3. (4) "AFF Makes Its Postwar Report," *AID*, December 1949, pp. 23–32.

[20] War Department, *AGF Light Aviation*, pp. 95, 98, 105, 111, 115, 119.

[21] "The All-Service Helicopter," *AID*, June 1949, p. 18.

[22] (1) Department of War, Kenneth C. Royall, "Report of the Under Secretary of War to the Secretary of War for the period of 1 July 1945 to 30 June 1947," Washington, DC, 22 July 1947. (2) General Dwight D. Eisenhower, *Final Report of the Chief of Staff, United States Army, to the Secretary of the Army* (Washington, DC: GPO, 7 February 1948), pp. 1, 4–6. (3) Louis Johnson, Secretary of Defense, "Teamwork Is Our Strength," *AID*, August 1949, pp. 19–24.

[23] (1) United States Army, Center of Military History, James E. Hewes, Jr, *From Root to McNamara: Army Organization and Administration, 1900–1963* (Washington, DC: GPO, 1975), pp. 243–44. (2) Weigley, p. 496. (3) OCAFF, "Report of the Army Equipment Board," pp. 2–3, 10.

[24] (1) "Is US to Have Three Air Forces? Worried Army Wants Own Support Planes," *US News & World Report*, 10 June 1949, pp. 20–21. (2) Hanson W. Baldwin, "Air Force Objective," *New York Times*, 27 June 1949, p. 12. (3) General Jacob L. Devers (Retired), "The Army Needs a Support Plane," *Flying*, March 1950, pp. 11, 58. (4) General J. Lawton Collins, Chief of Staff, United States Army, "Modern Weapons for Today's Army," *AID*, June 1950, pp. 3–9.

[25] "Digest of Speeches," *AID*, August 1949, pp. 56–61.

[26] "Digest of Speeches," *AID*, January 1950, p. 63.

[27] (1) United States Marine Corps, Headquarters, History and Museums Division, Lieutenant Colonel William R. Fails, *Marines and Helicopters, 1962–1973* (Washington, DC: GPO, 1978), p. 20. (2) Lynn Montross, *Cavalry of the Sky: The Story of US Marine Combat Helicopters* (New York: Harper & Brothers, 1954), pp. 44–47. See also foreword by Commandant of the Marine Corps, General Lemuel C. Sheperd, Jr. (3) Major Archie J. Clapp, United States Marine Corps, S3, Marine Composite Squadron One, "Their Mission is Mobility," *MR*, August 1953, pp. 10–18. (4) Galvin, p. 254.

[28] (1) Richard P. Weinert, Historical Office, Office of the Deputy Chief of Staff for Military Operations and Reserve Forces, US Continental Army Command, "A History of Army Aviation 1950–1962, Phase I: 1950–1955," Fort Monroe, Virginia, June 1971, p. 17. (2) Department of the Army, Office of the Chief of Transportation, "Army Helicopters, Transport and Cargo: An Evaluation of 1950–1953 Experience as a Basis for Current and Future Programming," Washington, DC, 1954, no page numbers, [pp. 1–3]. Hereafter cited as OCT, "Army Helicopters."

[29] United States Army, Airborne Panel, "Report of the Army Airborne Panel," 21 April 1950, Enclosure E, "Air Transportation," p. 1. Hereafter cited as 1950 Airborne Panel Report.

[30] 1950 Airborne Panel Report, pp. 4, 6.

[31] 1950 Airborne Panel Report, Enclosure E, pp. 2–3.

[32] See various memorandums within the 1950 Airborne Panel Report.

[33] The Army held an Army Airborne Conference on 19–23 February 1951 to disseminate the findings of the 1950 Army Airborne Panel. See United States Army, Office, Chief of Army Field Forces, "The Report of the Army Airborne Conference, 19–23 February 1951," Army Airborne Center, Fort Bragg, North Carolina, 17 March 1951, p. 4, Appendix A1 and C5.

[34] "The Structure of Army Aviation," *AW*, 14 April 1952, p. 83.

[35] (1) Tierney, pp. 180–81, 240, 243. (2) Major G. W. Kilmer, "Aircraft for the Army," *AID*, June 1952, pp. 38–44. (3) "Army Aviation Equipment," pp. 34, 37, 42, 44.

[36] House of Representatives, *Army Appropriations FY 54*, pp. 14–15.

[37] House of Representatives, *Army Appropriations FY 53, Part 1*, pp. 15–16.

[38] (1) Tierney, pp. 206–8, 212. (2) Montross, pp. 114, 118. (3) Lieutenant Colonel David M. Lam, "From Balloon to Black Hawk, Part III: Korea," *USAAD*, August 1981, pp. 44–48.

[39] House of Representatives, *Army Appropriations FY 54*, pp. 14–15.

[40] Department of Defense, *Semiannual Report of the Secretary of Defense and the Semiannual Reports of the Secretary of the Army, Secretary of the Navy, Secretary of the Air Force, January 1 to June 30, 1953* (Washington, DC: GPO, 1953), p. 115. Hereafter cited as DoD, *Semiannual Reports (date)*.

[41] (1) House of Representatives, *Army Appropriations FY 54*, pp. 14–15. (2) Laurence B. Epstein, "Army Aviation Logistics in Evolution, 1942–1953, Part 2," *USAAD*, July 1979, p. 35. Hereafter cited as Epstein 2.

[42] See Lieutenant Colonel Spurgeon H. Neel, Jr, "Aeromedical Evacuation," *Army*, April 1956, pp. 30–33.

[43] House of Representatives, *Army Appropriations FY 54*, p. 15.

[44] (1) OCT, "Army Helicopters," [pp. 1–3]. (2) Department of the Army, Transportation Corps, "Army Aircraft: Procurement and Production Programs," February 1953, p. 115. Hereafter cited as TC, "Army A/c P&P Programs."

[45] United States Army, Transportation School, Special Text ST-55-21-1, "Transportation Helicopter units," Fort Eustis, Virginia, June 1952. See also FM 20-100 before June 1952.

[46] (1) Captain Thomas L. McClellan, "Operation Bumblebee—How the US Marine Corps Developed Airmobile Tactics During the Korean War," *USAAD*, June 1988, p. 43. (2) Major Robert Winkler, "Height of Battle," *Infantry School Quarterly*, January 1955, p. 99. (3) Montross, pp. 134–35, 156–200. (4) United States Air Force, Tactical Air Command, "History of the Tactical Air Command, 1 January Through 30 June 1952, Volume V – Plans & Operations," Langley AFB, Virginia, n.d., pp. 28–29.

[47] (1) Lieutenant General John J. Tolson, *Airmobility 1961–1971* (Washington, DC: GPO, 1973), p. 4. (2) TC, "Army A/c P&P Programs," p. 121. (3) OCT, "Army Helicopters," [pp. 6–7], Annex C to P.

[48] (1) Senate, *DoD Appropriations FY 54, Part 1*, p. 886. (2) Department of the Army, Army Equipment Board, "Army Equipment Development Guide," Washington 25, DC, 29 December 1950, pp. 90–91. (3) DoD, *Semiannual Report, Jan.–June 1954,* p. 122.

[49] (1) Department of the Army, Office of the Chief of Staff, Washington 25, DC, Memorandum for Chairman, Materiel Requirements Review Panel, "Army Helicopter Program," 30 June 1952. (2) Department of the Army, Materiel Requirements Review Panel, Washington 25, DC, Memorandum for the Chief of Staff, US Army, "Materiel Requirements Review Panel Review of the Army Helicopter Program," 31 July 1952. Hereafter cited as MRRP, "Review of Army Helicopter Program."

[50] (1) TC, "Army A/c P&P Programs," pp. 7, 120, 122–26. (2) Williams, "Oral History," tape 2, pp. 37, 40.

[51] (1) Weigley, p. 569. (2) National Military Establishment, *First Report of the Secretary of Defense* (Washington, 1948), pp. 60, 141. (3) Department of Defense, *Second Report of the Secretary of Defense and the Annual Reports of the Secretary of the Army, Secretary of the Navy, Secretary of the Air Force, for Fiscal Year 1949* (Washington, 1950), pp. 140–41, 318. (4) DoD, *Semiannual Reports, Jan.–June, 1950* to *Jan.–June, 1955*, and *July–Dec., 1955*, pp. 3, 206; 103; 112; 99; 73–74, 332; 295; 2 respectively.

[52] See also Department of Defense, *Annual Report of the Secretary of Defense and the Annual Reports of the Secretary of the Army, Secretary of the Navy, Secretary of the Air Force, July 1, 1958, to June 30, 1959* (Washington, DC: GPO, 1960), p. 30. Hereafter cited as DoD, *Annual Reports FY (year)*.

[53] Department of the Army, Army Equipment Board, "Army Equipment Development Guide, Changes No. 1," Washington 25, DC, 17 December 1952, p. 3. Hereafter cited as DA, "Army Equipment Development Guide, Changes No. 1."

[54] (1) TC, "Army A/c P&P Programs," pp. 1, 41–44. (2) Epstein 2, p. 35. (3) US Army Military History Institute, Senior Officer Oral History Program, The History of Army Aviation, Interview of Colonel Delbert Bristol, US Army, Retired, by Colonel Ralph J. Powell and Lieutenant Colonel Ronald K. Andreson (Carlisle Barracks, Pennsylvania, 1978), tape 1, pp. 22–23. Hereafter cited as Bristol, "Oral History." (4) Williams, "Army-AF Conflicts," p. 3, and Tab O. (5) Major General Francis H. Griswold, Assistant Deputy Chief of Staff, Materiel, Department of the Air Force, Memorandum for the Chief of Ordnance, US Army, "Helicopter Procurement for the US Army," 1 November 1950, in MRRP, "Review of Army Helicopter Program," Staff Study 11.

[55] United States Air Force, Tactical Air Command, complied by historians Benjamin Goldman and Rex M. Naylor, and Major G. W. Hueners, Director of Historical Services, "History of the Tactical Air Command, 1 July Through 30 November 1950, Volume 1," n.d., pp. 286, 311–12. See also "Volume V," documents no. 832, 833. Hereafter cited as TAC, "History July–Nov. 50."

[56] (1) United States Air Force, Office of the Secretary of the Air Force, Special Projects Office, "History of the Office of the Secretary of the Air Force from July 1, 1951 to December 31, 1951, Volume I," n.d., pp. 29–30. Hereafter cited as OSAF, "History July–Dec. 51." (2) R. Earl McClendon, Air University Documentary Research Study AU-239-53-RSI, Documentary Research Division, Research Studies Institute, Air University, *Army Aviation, 1947–1954* (Maxwell Air Force Base, Alabama, 1954), p. 21. Pace was Army Secretary from 12 April 1950 to 20 January 1953.

[57] John A. McCone, Under Secretary of the Air Force, Memorandum for File, "Discussion with Secretary Pace, General Collins, General Hull, Army; Under Secretary McCone and General Vandenberg, AF; Subject: Liaison Aircraft Helicopters," 16 February 1951.

[58] (1) TAC, "History July–Nov. 50," pp. 286–87. (2) McClendon, pp. 21–22. (3) Williams, "Army-AF Conflicts," Tab O.

[59] (1) OSAF, "History July–Dec. 51," pp. 31–32. (2) John A. McCone, Acting Secretary of the Air Force, Memorandum for Secretary Pace, 1 August 1951.

[60] (1) "AF Yields on Army Plane Weight," *AW*, 5 November 1951, p. 17. (2) "Army Sets New Plans to Operate Aircraft," *AW*, 10 December 1951, pp. 13–14.

[61] (1) Ben S. Lee, "Army Due to Buy More Planes," *AW*, 13 November 1950, pp. 14–15. (2) "The Army's Role in Air Power," *AW*, 26 February 1951, p. 17. (3) "Army Copters," *AW*, 9 April 1951, pp. 17–18. (4) "Army Plane Purchases to Grow," *AW*, 16 April 1951, p. 15.

[62] The Evaluation Staff, Air War College, Air University, "Tactical Air Forces and National Security," Maxwell Air Force Base, Alabama, April 1955, pp. 23–31.

[63] Ben S. Lee, "Army Demands Plane-Buying Power," *AW*, 2 October 1950, pp. 12–13.

[64] (1) Williams, "Army-AF Conflicts," Tabs O, Q, R, S, T. (2) McClendon, pp. 24–25. (3) United States Air Force, Tactical Air Command, "History of the Tactical Air Command, Joint Army–Air Force Exercises Southern Pines and Long Horn, 1951–1952, Volume V: 1 July Thru 31 Dec. 1954," Langley AFB, Virginia, n.d., pp. 7–9, 23–29. (4) United States Air Force, Tactical Air Command, "History of the Tactical Air Command, 1 January Through 30 June 1952, Volume VI—Special Subjects & Activities," Langley AFB, Virginia, n.d., pp. 80–85.

[65] (1) Williams, "Army-AF Conflicts," Tab U. (2) MRRP, "Review of Army Helicopter Program," and Staff Study 1 therein.

[66] (1) Williams, "Army-AF Conflicts," p. 5, Tab V, Tab V Inclusion 1. (2) McClendon, pp. 25–26. Lovett was Secretary of Defense from 17 September 1951 to 20 January 1953.

[67] It is also possible that an incident involving an Army casualty evacuated by air ambulances led to this action. See: (1) Williams, "Oral History," tape 1, pp. 36–38. (2) United States Senate, Hearings Before the Special Subcommittee on Close Air Support of the Preparedness Investigating Subcommittee of the Committee on Armed Services, *Close Air*

Support (Washington, DC: GPO, 1972), 92nd Congress, 1st Session, pp. 99–100. (3) Williams, "Army-AF Conflicts," Tab V Inclusion 1. (4) OCT, "Army Helicopters," Annex E.

68 (1) McClendon, pp. 25–26. (2) Lieutenant Colonel O. G. Goodhand, Office of the Director of Army Aviation, Presentation at Aviation Conference, "Army Aviation Policy," 15 November 1956, pp. 4–5. (3) "Aircraft Accord Reached Between Army and USAF," *ANAFJ*, 22 November 1952, pp. 340, 361.

69 "Army Building Up Own Airlift Force," *New York Times*, 16 November 1952, p. 35.

70 Williams, "Army-AF Conflicts," Tabs X, Y.

71 (1) Harry S. Pack, Vice-President, Piasecki Helicopter Corporation, *A Survey of Helicopter Operations, Maintenance, and Supply in Korea* (Chevy Chase, Maryland: Operations Research Office, The Johns Hopkins University, June 1954), Appendix A, pp. 19–36. (2) Tierney, pp. 182–85. (3) OCT, "Army Helicopters," Annex Q: Comments.

72 General J. Lawton Collins, "New Approaches to World Peace," *AID*, January 1951, extract printed in OCT, "Army Helicopters," Annex Q: Comments.

73 Colonel Robert B. Neely, Assistant Chief of Transportation (Army Aviation), "Army Aviation," *ANAFJ*, 6 November 1954, p. 302.

74 House of Representatives, *Army Appropriations FY 53 Part 1*, p. 15.

75 Extract printed in OCT, "Army Helicopters," Annex Q: Comments.

76 (1) OCT, "Army Helicopters," Annex Q, Comments. (2) TC, "Army A/c P&P Program," p. 118.

77 McClendon, p. 24.

78 General Maxwell D. Taylor, Commanding, Headquarters, Eighth United States Army, Letter to Brigadier General Paul F. Yount, Acting Chief of Transportation, Department of the Army, Washington 25, DC, 1 October 1953. Reprinted in OCT, "Army Helicopters."

79 (1) Montross, p. 221. (2) OCT, "Army Helicopters," Annex Q: Comments.

80 OCT, "Army Helicopters," [introduction].

81 For comments by Army Under Secretary Karl E. Bendetson see "Sikorsky's in the Army Now!" *Bee-Hive*, Fall 1952, pp. 16–17.

82 Frank Pace, Jr, Secretary of the Army, Address Before the AH Society, Washington Hotel, Washington, DC, "Air Mobility for the New Army," 15 May 1952. Extract printed in OCT, "Army Helicopters," Annex Q: Comments.

83 Earl D. Johnson, Under Secretary of the Army, no title, *NDTJ*, September–October 1953, p. 64.

84 Earl D. Johnson, Under Secretary of the Army, Address Before the Dallas Council on World Affairs, Dallas, Texas, "Symposium on Air Power," 20 November 1953. Extract printed in OCT, "Army Helicopters," Annex Q: Comments.

SOURCES FOR TABLES

2-1. (1) Bureau of the Budget, *Budget of the United States Government for Fiscal Year 1950* and *1951* (Washington, DC: GPO, 1949 and 1950). Hereafter cited as Bureau of Budget, *FY (date)*. (2) House of Representatives, *Army Appropriations FY 50*, *FY 51*, and *FY 52*, pp. 427, 744, 525 respectively.

2-2. TC, "Army A/c P&P Programs," pp. 116–19.

2-3. McClendon, p. 29. Figures exclusive of aircraft assigned to the National Guard.

2-4. MRRP, "Review of Army Helicopter Program," Staff Study 3, Tab C.

2-5. (1) Bureau of Budget, *FY 53*, *FY 54*, and *FY 55*. (2) House of Representatives, *Army Appropriations FY 53, Part 2*, *FY 54* and *FY 55 [Part 4]*, pp. 776, 1469, 1035 respectively. (3) Jerry Greene, "Army Aviation," *Pegasus*, October 1953, p. 7. (4) "New Moves Boosts Army Aviation Growth," *AA*, Vol. 16, No. 14 (8 December 1952), pp. 17–18.

2-6. Williams, "Army-AF Conflicts," Tabs P, W.

II. The Eisenhower Period

3

AIR MOBILITY RATIONALES UNDER THE "NEW LOOK" POLICY

Under the influence of the Eisenhower administration's national security policies, several new rationales for air mobility emerged from 1953 to 1960. This included the need for more mobility, both air and ground, on the nuclear battlefield, and more strategic and tactical mobility for overcoming the strategic inferiority in ground forces. Near the end of this period, limited war took on a more important political profile. Air mobility for limited wars accordingly took on an increased importance. Apart from administration policies the lessons of Korea and the inherent military logic of transport helicopters were also providing an impetus for developing air mobility. External stimulus such as foreign developments of helicopters and air mobility, on the other hand, was not a very important factor. The same was true of interservice rivalry with the Air Force. The lessons of Korea and the existence of nuclear weapons also led the Army to develop a new concept distinct from air mobility called sky cavalry.

THE EISENHOWER ADMINISTRATION'S MILITARY POLICY

The central tenet of the Eisenhower administration's New Look policy was to balance economic and military needs. It stressed sustainable military expenditures over the long term. The military side of the New Look policy was in essence an asymmetrical response to the Soviet threat. It placed greater relative emphasis on nuclear weapons and strategic air power, even though the administration reduced the Air Force's force structure goals. It also relied on other nonmilitary components (such as alliances, covert action, psychological warfare, and negotiations), rather than conventional forces. The administration wanted its allies or indigenous armies to provide the bulk of the ground forces and to do most of the ground fighting. The United States might supply technical assistance or intervene with air or naval forces if necessary. This would reduce costs and the fiscal pressure on the government budget, and reduce the burden on the economy (balanced budget, lower taxes, less borrowing, less inflation).[1] The effect of the asymmetrical policy was a reduction in the number of personnel. This policy affected the Army the most, since it had an almost complete orientation towards conventional war-fighting. Its traditional role suffered and Army morale as a whole deteriorated.[2]

While administration policies discriminated against large conventional ground forces, this did not imply that ground combat was obsolete. To offset the reduction in conventional ground forces the administration policy was to use tactical nuclear weapons to deal with conventional general war, or large limited wars (Korea-type). Moreover, there were exceptions to the administration's asymmetrical policies. One was the stationing of significant ground forces in Europe as a deterrent to Soviet ground attack. The administration also maintained forces in Korea to deter another North Korean attack. This was a radical departure from traditional United States foreign policy. It also wanted to create a mobile, central strategic reserve of six Army divisions, presumably to reinforce Europe, Korea, or trouble spots elsewhere.[3] This directly contradicted the "no intervention" policy.

Contradiction notwithstanding, two Army Chiefs of Staff, Ridgway and Taylor, opposed the administration's policy of assigning low priority to limited war programs and attempted to justify or maintain the Army's relevance.[4] They also embraced tactical nuclear weapons as a way of jumping on the nuclear bandwagon.[5] The Army adapted its doctrine to meet these changing policies. The defense of Europe was the Army's primary concern, followed by limited war functions.[6] Air mobility was a central tenet in each of these policies. Since Europe was vital to United States interests, a Soviet attack would elicit a response with American tactical nuclear weapons. However, it was possible that the Soviet Union possessed tactical nuclear weapons as well.[7] The Army, having to deal with this contingency, began to reorient itself to fight on a nuclear battlefield. The threat of nuclear attack presented problems that only more mobility could solve, in the form of air mobility and ground mechanization. The Army also made its divisions strategically more mobile and easier to deploy by air to fulfill the role of the central reserve. This also had repercussions for battlefield mobility.

TACTICAL NUCLEAR WEAPONS

Ridgway and the Transition to a Nuclear Battlefield

The first official statement regarding the role of transport helicopters in nuclear warfare appeared in the 25 November 1952 Army Equipment Development Guide. This simply stated three purposes of Army aviation. First, it was for "expediting and facilitating the conduct of operations on land." Second, it was for "improving mobility, command, control, and logistic support of Army forces." Third, it was for "facilitating greater battlefield dispersion and maneuverability under conditions of atomic warfare."[8] In 1953 the Acting Chief of Transportation, Brigadier General Paul Yount, expanded on the transport helicopter's role. He listed four missions:

> a. Logistical support of Army troops when enemy operations, terrain conditions, or their own advance has severed their normal supply system;
> b. Tactical operations with small reserve and maneuver elements to maintain the battlefield mobility required to counteract the enemy's operations;
> c. Logistical support of tactical special weapons to preserve their security and surprise capabilities;
> d. Tactical operations in conjunction with new special, high destructive [*sic*] weapons to exploit shock and disorganization of the enemy forces.[9]

Brigadier General Yount gave one specific example of transport helicopter operations under nuclear conditions. He pointed out that transport helicopters offered the only swift method of exploiting a nuclear strike on the enemy:

> Exploitation of the effects of these [nuclear] devices demands a prompt follow-up of their very high shock potential by promptly moving troops into the area attacked before the enemy can reform and reorganize his forces. The helicopter with its ability to overpass destroyed roads and bridges and deliver troops into an

> area 20 or 30 miles away in a matter of minutes, offers the only solution to this problem.[10]

Soviet tactical nuclear weapons were similarly a threat to the Army's own ground operations. In March 1955, Major General James Gavin told an interviewer that "the important thing is to realize that the Russian has the same atomic weapons we have." He added that "technologically and tactically he can do to us what we can do to him."[11] Gavin was then the Assistant Chief of Staff for Operations, G-3 (or simply known as G-3), an important post in the Army high command. The Army therefore began to reorient itself to adopt a nuclear war-fighting doctrine for both offensive and defensive operations.

After General Ridgway became the Chief of Staff on 16 August 1953, he directed the Army to undertake a radical review of land warfare. Led by General Gavin, this was to study the effects of nuclear weapons on ground operations. Ridgway told those deliberating on the review to cast aside their conservatism and reliance on the tried and tested.[12] Even Gavin himself said: "We generals . . . are always accused of trying to fight the next war with the weapons of the last. This time we're trying to get ready to fight the next war, if it comes, with the weapons and organization of the next."[13] The Army began to contemplate and experiment with new field organizations,[14] including new divisional aviation organizations.[15]

Ridgway and the Army staff concluded that the superior firepower offered by nuclear weapons would not mean much unless the Army could feed, supply, and transport its combat units quickly. They said the Army must downgrade the importance of fixed lines of communication and render huge supply dumps obsolete. They concluded that their vulnerability to nuclear attack meant that aircraft must provide the means for troop transport, resupply, evacuation, and communications to a far greater extent than before. This was especially true, Ridgway stated, when there were swift changes in the location of widely scattered combat units.[16] The objective of Army helicopters was to be able to move 20 percent of the Army's front line transportation requirements.[17] General Gavin said that "the whole character of modern war presumes a basic dependence on air vehicles. The key to atomic warfare tactically will be how to keep isolated units alive over long periods, still fighting when they are deeply encircled behind enemy lines."[18]

Taylor and the Pentomic Reorganization

Ridgway's review did not lead to any major Army-wide doctrinal upheaval within his tenancy as Chief of Staff. This was due to the complicated nature of tactical nuclear warfare and Ridgway's short term of office. When General Taylor became Chief of Staff on 30 June 1955, he continued the work started by Ridgway. The Army's major command in the United States, the Continental Army Command or CONARC, prepared a new study from September 1955 known as the Pentagonal Atomic-Nonatomic Army (PENTANA). This study developed organizational and doctrinal concepts for the field army in the 1960–1970 period. It became the basis for the later Pentomic reorganization. Exercise SAGE BRUSH (or SAGEBRUSH), held from 31 October to 15 December 1955, put these new concepts for the nuclear battlefield to the test. After the exercise, Taylor incorporated the nuclear dimension in Army combat units and major doctrinal changes occurred.[19] The nuclear rationale grew so pervasive that limited war even became understood as limited nuclear war. Undoubtedly, under Taylor the nuclear rationale became the prime impetus for the development of air mobility. This was evident from articles by Secretary of the Army Wilber Brucker and Army officers in the staff and planning organizations.[20] It was also evident in Taylor's deeds and speeches.[21] He told Congress in 1958:

> The basic dilemma facing all military forces—ours and the Soviet's alike—arises out of the fact that our ability to deliver nuclear firepower has outstripped our efforts to reduce our own vulnerability to those same nuclear fires. In fact, we can never reduce our vulnerability to atomic fires to a satisfactory extent without a decisive jump in mobility. We seem to have gone just about as far as we can go in the research for mobility through improved ground vehicles. There will be improvements in the future, of course, in range, flotation, combat worthiness, and armament. However, none of these improvements in ground vehicles will overcome the traditional barriers of terrain to the extent that they must be overcome if we are to take the large step forward required by nuclear warfare. We must look for aerial vehicles which can move across the terrain at 60 or 70 miles an hour, whether that terrain consists of mountains, jungles, swamps, rivers, or plains. This is the direction in

> which Army aviation must lead which, I reiterate, is not in the direction of another air force.[22]

Taylor pointed out five problems inherent in tactical nuclear warfare to which better mobility was a solution. They were: unit survival, target acquisition, protection from radiological fallout, avoiding enemy detection, and exploiting Army firepower.[23]

The characteristics of a nuclear battlefield dictated that Army subunits within the division must be able to fight dispersed and in all directions. In 1956–1957 Taylor reorganized Army units for omnidirectional combat based on new organizational structures (the so-called Pentomic reorganization). The Army abandoned the traditional triangular structure and adopted the pentagonal structure. The Army designated the new divisions using the names Reorganization Of The Airborne Division (ROTAD), Reorganization Of the Combat Infantry Division (ROCID), and Reorganization Of the Combat Armored Division (ROCAD). Each infantry or airborne division had five battle groups, while each battle group had five companies. The battle group replaced the battalion and regiment or brigade level of command. Only the armored divisions retained their combat command structure as developed in World War II. Each battle group was a self-contained force capable of semi-independent combat. There was also a substantial reduction of manpower and unnecessary heavy equipment. The new airborne division had 11,486 men compared to 17,085 formerly. The new infantry division had 13,748 men as compared to 17,455 formerly. This made the ROTAD completely air transportable, while improving the air transportability of ROCID.[24]

The central idea behind the Pentomic concept was to enable units to deploy in a checkerboard fashion, thus presenting no lucrative nuclear targets to the enemy. For this concept to work, the Army required strong firepower to defeat overwhelming enemy forces (mainly in the form of battlefield nuclear weapons) and organic aviation for target acquisition. Army forces also required mobility to

Table 3-1
Aircraft in Pentomic Divisions

	Airplanes		Helicopters		
Divisions:	Observation	Utility	Recon	Utility	Light Cargo
Infantry	16	6	20	8	–
Armored	19	5	12	8	6
Airborne	10	6	17	–	20

Table 3-2
Division Aviation Company (ROCID)

	Aircraft	Comments
Direct Support Platoon:		
• Artillery Flight	10 x L-19	
• Target Acquisition Section	3 x L-19 3 x L-20	Equipped with surveillance devices. To team up with the reconnaissance and surveillance platoon of the divisional cavalry squadron.
• 5 x Combat Flights	5 x H-13	One flight to support each battle group as necessary.
General Support Platoon:		
• Command Support Section	3 x L-19 3 x L-20	Normally used by the division commander and his staff officers for reconnaissance and liaison.
• Tactical Support Section	15 x H-13	For reinforcing the combat support flights or other units of the division as required.
• Utility Section	8 x H-19	Capability to move troops and supplies. Under normal conditions this would be 48 troops or 6 tons of supplies in a single lift.
Service Platoon.		

fight dispersed and then be able to concentrate rapidly for attacks (with highly mobile ground and air vehicles). Forces had to redisperse rapidly again after the attack to present a lucrative target to enemy nuclear weapons for as brief an opportunity as possible. Lastly, the Army needed the communications for controlling the forces.[25] Taylor and the Army high command thus strove to improve the battlefield mobility of dispersed ground forces, both in the air and on the ground.[26]

Taylor's reorganization created a divisional aviation company and almost doubled the number of aircraft per division. Previously, each Korean-era division had 26 or 28 aircraft. They were mostly observation airplanes and reconnaissance helicopters with only 3 utility airplanes (L-17 or L-20). The Pentomic reorganization increased the number of divisional aircraft to 50 (53 for airborne). Within this there was a substantial increase in the number of light cargo helicopters, utility airplanes, and utility helicopters (see Table 3-1). Each aviation company had one direct support platoon, one general support platoon, and one service platoon (see Table 3-2 for an example of the ROCID aviation company). This was a significant increase in divisional aviation capabilities without sacrificing aviation capabilities at the field army level. This addition allowed divisions to have a limited capacity to perform aerial logistical support missions and tactical operations with less dependency on external aviation assets.

STRATEGIC MANPOWER INFERIORITY

While the nuclear rationale provided the prime impetus for air mobility, this was not the exclusive impetus. The Army also felt it needed air mobility to overcome its strategically inferior size compared to the Communist Bloc army.[27] In 1953 Brigadier General Yount felt that "our inferior forces will be forced to fight against the enemy's masses in the event of a future war."[28] In December 1954 Ridgway said:

> In its military meaning, mobility can be defined as the ability to shift striking power. In a sense, the effective strength of a military force—the strength which can be applied against the enemy—is multiplied many times by mobility. Combined with proper use of the principles of the objective, of deception, security, surprise, offense, and economy of force, it can give local numerical superiority over an enemy whose over-all strength is greater than our own. Its importance, therefore, in any conflict between us and enemy masses which we might have to face, is strikingly obvious, as would be our peril, should our mobility not exceed that of an enemy of that kind.[29]

This assessment originated from the tendency to compare the balance of forces between the United States and the communist bloc of China and the Soviet Union, without including allies in the calculations. In 1952 an officer compared the American population of 150 million with the Soviet Union's 205 million and its satellites' 575 million people. The Soviet Union alone could raise 300 ground divisions compared to America's maximum of 100.[30] However, the Army did not always take such a one-sided calculation. In 1957 Secretary Brucker noted that there were over 200 allied divisions within the "Free World," which included NATO and SEATO.[31] Nevertheless, in June 1958 an article in the *Army Information Digest* stated:

> Numerical inferiority in a non-atomic limited war must be offset by attaining and exploiting any significant US Army superiority in conventional

> firepower, ground and air mobility, control means, and imaginative tactics. The field of mobility, if exploited, offers opportunities for strategic and tactical innovations comparable to those made possible by atomic-bearing missiles.[32]

The Army also looked toward air mobility to bridge the gap in combat capability between its personnel reduction and its military mission. The post-Korea reduction in Army personnel meant it could not have permanent deployments around the world's trouble spots. It therefore recognized the need for a global "fire-brigade" capability (or a strategic cavalry force) to respond to or to prevent future Korea-style crises or small "brush fire" wars. Ridgway and the Army staff's discussion of air mobility included both tactical and strategic dimensions.[33] Ridgway said:

> Mobility exists in a number of forms. There is tactical mobility—the ability to shift striking power on the battlefield. And there is strategic mobility, which in its simplest form, might be described as the ability to shift striking force *to* the battlefield, or from one battlefield to another battlefield.[34]

To Ridgway, the need for strategic deployment and in-theater mobility meant the Army must possess long-range and short-range mobility far beyond anything ever known in war before. In his 1956 book, he said the future Army must be greatly dependent upon aircraft of one form or another. He said the Army required long-range transport aircraft for intercontinental movement and a number of specialized aircraft for battlefield employment. These battlefield aircraft need not be long-range or high-speed planes, but in essence were aerial trucks and jeeps for transport, and flying gun platforms for combat.[35] Air-transportability of Army equipment and weapons became the most important requirement to fulfil this strategic cavalry role.[36] However, the achievement of this goal was difficult because sustained land combat in the nuclear era was requiring heavier and bulkier equipment.[37] Taylor, who was an airborne officer during World War II as Ridgway was, continued to emphasize the need for a strategic air transportable force.[38] In October 1955 Taylor said:

> Although the expense of this kind of movement may seem great, strategic air mobility multiplies the effectiveness of our central reserve and reduces the

> number of troops which must be kept overseas. Thus it can be, in fact, a great economy. . . .
> In short, the Army today is bent on reaching a condition of airborne effectiveness which we are convinced is necessary for the successful conduct of modern warfare. We want an Army with many completely air-transportable combat units, complete with sufficient weapons, vehicles and supplies to sustain themselves in ground operations.[39]

The manifestation of this lay in the formation of the Strategic Army Corps (STRAC) with four divisions in 1958. However, because of personnel reductions only three divisions were available in 1959.[40] Moreover, the strategic mobility plan required adequate Air Force airlift to work. What constituted adequate airlift was a continuous source of contention between the Army and the Air Force.

Besides tackling outbreaks of aggression, the Army felt that a strategically mobile force could preempt or deter or even defuse a crisis, in the place of permanently stationed ground forces. Secretary Brucker's view was that the Army needed strategic and tactical airlift to come to the support of allied nations quickly in time of crisis. The Army saw this capability as providing a strong deterrent to local aggression.[41]

Taylor summed up the need for mobility to deal with both distant crises and numerical inferiority:

> The ability to execute movements of all kinds over short and long distances constitutes mobility—the attribute which a successful Army must have to give effect to its weapons. It provides the speed differential over a slower antagonist which allows the possessor to place destructive firepower at the right time and place to do the most hurt to the enemy.
> Thus, in the military sphere, mobility may be defined as the ability to shift firepower. It multiples the effective strength of the military force which can be applied against an enemy without increasing the available means. It can give local superiority over an enemy whose over-all strength is greater than our own. It permits the rapid shifting of reserves to meet sudden crises in distant areas. Thus mobility is a form of economy which our Army can not afford to lack.[42]

The new Army Chief of Staff, General Lyman Lemnitzer, who took over from Taylor on 30 June 1959, also believed in the objective of attaining an improved strategic mobility.[43] Under Lemnitzer, considerations of strategic inferiority, inadequate forces for permanent overseas deployments, plus the need to have a fire brigade force were still predominant.[44]

EXTERNAL STIMULUS

A comparable transport helicopter development in the Soviet armed forces during the fifties achieved a certain degree of publicity.[45] However, there is scant evidence that this influenced the Army high command. The Army Secretaries and the Chiefs of Staff seldom mentioned Soviet developments in either their speeches or congressional testimony. Official Army aviation plans also did not mention Soviet developments. The Army high command's investment in air mobility was therefore not a reaction to developments in the Soviet Union. This does not mean, however, that lower-ranking officers such as Army aviation leaders were ignorant of Soviet developments. Indeed, some officers were using Soviet developments to justify the Army's own. This was especially true in the late fifties.[46]

In June 1959 the Director of Army Aviation, Brigadier General Clifton von Kann, urged the Army to utilize its aviation potential more effectively than potential enemies. He warned that "our enemies are not asleep to the possibilities in this field. So there is pressure upon us to develop our capabilities faster than he."[47]

In September 1959 Brigadier General Ernest Easterbrook, Commandant of the Army Aviation School, pointed out that the Soviet Army had a positive program for increasing its battlefield mobility with helicopter mobile forces. He believed that their intention was to have a fleet of 2,000 transport helicopters. He claimed that the Soviet Army recognized and was well aware of the helicopter's value in gaining tactical surprise and to provide operational mobility. He also stated that the Soviets were pursuing an active program of arming helicopters. Easterbrook said: "One thing appears certain, if our Army doesn't capitalize on the ability to utilize the air medium, others will take the lead."[48] Later Easterbrook speculated that the Soviets might form air mobile combat units for offensive missions, with several hundred transport helicopters and supporting armed helicopters. He therefore argued that the Army should have better surveillance aviation and air mobile task forces for dealing with the threat. He even foresaw a requirement for helicopter air-to-air combat.[49]

The Army aviation community also took note of helicopter usage in Malaya, Kenya, and Algeria.[50] This attention was a result of American equipment being in use. The British were using the S-55 (H-19) helicopter to combat guerrillas in Malaya. The French had a variety of helicopters in Algeria, including American-made transport helicopters. The French were using commando forces to attack rebels by helicopter, something that would interest the Army. In addition, the American manufacturers, who sent representatives out to Algeria, had a keen interest in the operations and maintenance of helicopters. They returned to make recommendations to the Army based on French experience.[51] The helicopter manufacturers even brought back French officers from Algeria to provide first-hand information on their helicopter operations.[52] The Army Aviation Center under Brigadier General Carl Hutton as well as other Army aviation authorities studied these technical, operational, and tactical recommendations.[53] The French also used armed helicopters in Algeria as escorts. In early 1958 the Commandant of the Army Aviation School, Brigadier General Bogardus Cairns, spoke of the French experience in Algeria and how mandatory it was to arm helicopters.[54]

LESSONS OF KOREA

It was apparent to the Army that, from an evolutionary standpoint, transport helicopters were ideal for augmenting ground lines of communication. Brigadier General Yount said in 1953:

> The cargo helicopter with the Army today represents one of the greatest single improvements to military transportation in history. In an effort to increase our battlefield mobility we have steadily increased the number, size and weight of our highway vehicles, and, in the main, the result has been to render us at least as susceptible as was Napoleon to poor roads, mud and snow. The very features of the helicopter which make it unattractive as an air transportation vehicle in many ways make it particularly applicable to our problem. Its short range of 75 to 150 miles appears infinite in the eyes of the foot soldier; its slow speed of 90 to 100 miles an hour is supersonic to the truck

> driver fighting through mud and snow at a snail's pace; its limited, two to three tons, capacity is adequate to supply food or ammunition for several hundred troops for a whole day.[55]

Korea also furnished important lessons on air mobility despite involving only 220 helicopters in total.[56] In April 1954 *Harper's Magazine* published a landmark article entitled "Cavalry, and I Don't Mean Horses!" by Major General James Gavin. This article was an unofficial summation of the studies done under Gavin while he was responsible for Army planning and operations as the Assistant Chief of Staff, G-3. Some of the studies consist of replays of the Korean war campaign operations using helicopters.[57] The article emphasized the use of completely air mobile units to perform tactical tasks, rather than mere transportation. Gavin expounded two theoretical justifications for this, one of which drew heavily on the lessons of Korea. He argued that the Army was in critical need of a force that could move more swiftly than the main army; or in his own term, a force with a mobility differential. This in essence was analogous to the mobility differential between horse cavalry and foot-marching armies. Gavin observed that by the start of the Korean war such a differential had ceased to exist in modern armies. Most modern armies possessed the same degree of motorization, while within each army there were no physical distinctions between "cavalry" and other forces. A main force moved at the same speed as a screening force, making it difficult for such a screening force to carry out its designated role. This anomaly was most acute during the Korean war, when the heavily equipped and road-bound American forces moved north after the Pusan breakout without adequate screening. Gavin felt that for this reason the Chinese forces surprised them.[58] Gavin also argued that there was a historical cycle in the ascendancy of firepower and mobility. He felt that firepower, in the form of nuclear weapons, had reached a very high crest at this point in history. The solution to redress the lack of mobility differential and the overwhelming ascendancy of firepower, Gavin stated, lay in the air vehicle and the airborne-armor combination.[59]

After its publication, the article acted as the clarion call to solve the problems of mobility differential.[60] The Army also used the concept of the firepower-mobility gap extensively to justify air mobility (see earlier discussions).[61] Eventually, the nature of a nuclear battlefield blended with mobility differential to act as a prime rationale for sky cavalry. With greater dispersion and more fluid operations, reconnaissance elements had several times the area to cover compared to the past. Highly mobile ground and air reconnaissance and surveillance elements screening division fronts would minimize infiltration, acquire targets, force the enemy into creating remunerative targets, and direct the employment of supporting weapons.[62] The Army considered the tank not fast enough to find targets for nuclear weapons. Moreover, impassable and rugged terrain would limit its use to roads (as in Korea). Aircraft, used the same way as horse cavalry during the American Civil War, were the answer. Besides giving additional capability, the Army felt sky cavalry units would decrease the need for men (or rather, there was going to be a lack of manpower for extensive ground reconnaissance and security).[63]

The Department of the Army approved the concept of "air mobility for land reconnaissance" in the March 1956 Army Aviation Plan (see Chapter 4 for more on this plan). It recognized the need for a concept that embodied Gavin's original rationale:

> The dispersion of units to be effected on the future battlefield, the highly fluid character of operations, and the mobility now characteristic of all tactical units in modern armies, makes desirable or even mandatory the development of "air cavalry" forces whose speed and flexibility of movement are enhanced by the use of aircraft. Thus may be provided the "mobility differential" essential to the units which must perform the missions classic to highly mobile light elements: reconnaissance, security of open flanks, seizure of critical areas, pursuit, and to some extent exploitation.[64]

LIMITED WARFARE

While during this period the Army never abandoned the belief that limited wars were likely to occur, its main focus was on the problems of nuclear warfare. At the end of Taylor's tenure the place of limited warfare began to grow in importance again, perhaps because of the Lebanon intervention in 1958. When Lemnitzer became Chief of Staff, the Army began a transition away from the all-pervasive nuclear rationale. The emphasis on anti-guerrilla warfare increased. However, the nuclear rationale still dominated Army policy, perhaps because of the Berlin crisis starting in 1958. In both nuclear and limited wars the Army saw air mobility as very important. The conceptual and matériel studies of the late Eisenhower period all had an increasing air mobility content.

Lemnitzer and the Transition to Limited War

The prominence of the nuclear rationale was still apparent in an article by Lieutenant General John Oakes, the Deputy Chief of Staff for Military Operations. He wrote in September 1960 that greater tactical mobility was the trend of the sixties. This was to solve the dilemma posed by the need for dispersion to avoid massive nuclear attack versus the need for sufficient concentration to accomplish a mission. The Army's aim was to have 100 percent vehicular mobility, ground or air, for all combat and combat support units. He pointed to the growth of Army aviation and ground mobility as evidence of progress in this area. Both forms of mobility, he said, were swinging the (nuclear) firepower-mobility pendulum back to mobility.[65] Other articles written by Army aviation leaders of this period also had the nuclear battlefield as the main rationale.[66] Furthermore, the service schools like the Infantry still saw the threat or actual use of nuclear weapons in their future scenario.[67] In Europe, an Army exercise in 1961 still emphasized limited nuclear war.[68]

However, there was a detectable shift to a limited war orientation as Taylor's tenure as Army Chief of Staff was ending. First, the practicality of the use of tactical nuclear weapons and the pentagonal structure came increasingly under question.[69] This was especially true since the concept required weapons and equipment beyond what was then available. Electioneering during the last years of the Eisenhower administration also generated political debate about the wisdom of relying solely on nuclear deterrence. The national security policy of massive retaliation looked increasingly untenable and a new administration was certain to change it, whichever party won.[70] Some officials within the Eisenhower administration also became dissatisfied with the nuclear strategy.[71] An increased emphasis on limited war developed within the Eisenhower administration. In September 1960 the JCS considered establishing a new specified command based on limited war forces.[72] This eventually became Strike Command or STRICOM, which combined air, ground, and naval forces. The increasing number of worldwide conflicts based on guerrilla warfare, added to the likelihood of United States intervention, made this necessary. Press reports in September 1960 stated that Lemnitzer believed a shift to limited war expenditure was necessary in all three services.[73]

Lemnitzer argued that the conduct of military operations in a nuclear war would be considerably different from nonnuclear war. He said that, in the main, nonnuclear ground warfare was substantially unchanged from World War II or Korea. This statement in itself indicated a change of emphasis. The Army had earlier emphasized that ground forces would still operate dispersed even if neither side was using battlefield nuclear weapons. The mere threat or possibility of their use dictated continued vigilance so as not to present a lucrative target. Otherwise it might tempt the enemy to initiate nuclear use.[74] Moreover, Lemnitzer tried to ease the Army into guerrilla warfare using the nuclear rationale. He compared the fundamental requirement of nuclear warfare, which greatly increased the dispersion of units, with the characteristics of guerrilla warfare:

> One of the consequences of increased dispersion is that infiltration by both sides must be expected. Ground combat would be conducted in the combat *zone*, rather than along a "front line," as in the past. . . . In many respects, battle would take on a number of the characteristics of guerrilla warfare. . . . All this has required us to develop new tactical doctrine, and to adopt new forms of tactical organization to carry out that doctrine. Guerrilla warfare is being conducted in several important areas in the world today.[75]

In another address, General Lemnitzer looked five years into the future and detailed the nature of the Communist threat that he foresaw. He spoke of general and limited wars, and then discussed guerrilla warfare:

> [W]e need to improve our capabilities and those of our Allies to conduct anti-guerrilla warfare. This is becoming an increasingly active area. In Southeast Asia and in other parts of the World we can expect to encounter guerrilla activities because the Communists are showing themselves very able in this field. Anti-guerrilla capabilities are necessary to permit situations such as exist today in Vietnam and Laos to be brought under control by those governments which are friendly to the United States.[76]

General Lemnitzer specified the necessary steps the Army needed to take for nuclear wars as well as other types of war, one of which was to improve air mobility:

> In addition to modernization of weapons, there is also the need for modernization in the vital field of tactical mobility. . . . There is no doubt . . . the expanded area of the battlefield emphasizes the importance of mobility. It is highly possible that the traditional restrictions upon land warfare imposed by terrain features such as rivers, swamps, mountains and forests can be surmounted through forthcoming developments, particularly with regard to the field of Army aviation and tactical air mobility.[77]

Similarly, the Director of Army Aviation, Brigadier General Clifton von Kann, mentioned in June 1959 the importance of Army aviation on a nuclear battlefield. However, he also acknowledged that the world had become locked in a missile stalemate. He therefore concluded in an address in July 1959 that "the critical thing now is our capability to fight a limited war."[78] By November 1960 von Kann made the argument that Europe was not the likeliest military contingency:

> Up to World War II our thinking was based largely on temperate zones, and much of our military gear is still designed for such climates. However, World War II, Korea, and the succession of brush-fire wars tell us that we may be called upon to fight anywhere between the Poles and quite possibly on the Equator.
>
> Now an analysis of ground trafficability, even if we restrict ourselves to the relatively favorable picture along the periphery of the Iron Curtain, demonstrates that a very substantial portion of this area prohibits the wide use of ground vehicles, and tremendous engineering effort would be necessary if they are to move at all. It is difficult to escape the conclusion that only an air-mobile force has true mobility in the Arctic muskeg or the virtually impassable jungle. The recent events in the Congo illustrate the point vividly. . . . [T]he key to success was held by that side which could control the airfields and the lifelines of air transport.[79]

Finding a Pentomic Replacement

The search for a replacement to the Pentomic division was not just a step taken to reorient the Army towards limited warfare. It was also because the Pentomic division had serious shortcomings. In January 1959, General Bruce Clark, the Commanding General of CONARC, ordered the preparation of an organization study called Modern Mobile Army 1965–70, otherwise known as MOMAR I or MOMAR-70. CONARC completed the first draft in July 1959 and submitted it for review by all the combat development agencies, the Army's service schools, and the six continental Army headquarters. A General Officers' Board then reviewed the final draft version from 15 October to 24 December 1959. CONARC published the final version in February 1960. While this study retained elements of the pentagonal structure, the main emphasis was on limited rather than general wars.[80]

The MOMAR study planned a considerable increase in Army aviation in both numbers and tactical organization. For example, there would be a General Support Aviation Company and a Reconnaissance and Security Squadron within each medium and heavy division. The Reconnaissance and Security Squadron would replace the conventional cavalry in existing divisions. It was a combined-arms unit completely mobile in organic air vehicles and would therefore have a mobility differential over other divisional elements. In addition, the medium division would have a Tactical Transport Company with the required numbers of transport aircraft to lift the essential elements of a task force from the Light Combat Command. (Each medium division would have three medium and two light combat commands, with the latter designed for participating in air mobile operations.) A MOMAR field army (there was no corps level) would have an Army Aviation Transportation Group, an Army Reconnaissance Group, and an Aerial Surveillance Battalion. The Army Reconnaissance Group would be another air mobile successor of the Armored Cavalry Group. The Army Aviation Transportation Group would have two fixed wing STOL "elements" and two rotary wing VTOL "elements." The VTOL elements would be sufficient for lifting two task forces plus equipment from a light combat command. In addition, the VTOL elements would have additional aircraft to assist in logistical support of forward elements, moving up to 25 percent of administrative tonnage, or 100 tons daily. The rotary wing aircraft would have a suppressive fire capability.[81]

The MOMAR study was important because it encapsulated many radical ideas, including armed helicopters and air cavalry units.

Helicopter-mounted, aerial weapon systems were to deliver suppressive fire on enemy positions along the flight route to and from the objective. They were also to provide additional fire support to the air mobile force in the objective area. Both fire missions were to overcome partially the lack of fire support that was inherent in an air mobile force, which did not have tanks, armored personnel carriers, or heavy artillery.

In April 1960 the Command and General Staff College became the coordinating agency for the continued development of the MOMAR concept. It was to fill in, expand, and refine the concept before the Department of the Army approved it. When the college completed the new study, it strongly emphasized that the Army had to be capable of operating in a wide variety of tactical circumstances. This ranged from a limited war without nuclear weapons, to a general war with nuclear weapons.

Meanwhile, the Department of the Army was not receptive to some of the ideas included in MOMAR I. A letter by the Vice Chief of Staff General Clyde Eddleman dated 16 December 1960 reflected this. At this point the threat of guerrilla warfare had not yet redirected Army thinking. However, it was clear that the heavily mechanized forces of MOMAR I were unsuitable for many areas of the world. The MOMAR organizational concept lacked the necessary flexibility to meet the Army's needs, and so General Eddleman shelved the project. He provided new guidelines and initiated a new study that eventually led to the ROAD reorganization (see Chapter 7).[82]

INTERSERVICE RIVALRY AS AN INCENTIVE

There is a possibility that interservice rivalry provided an impetus to air mobility developments. The Army had a grievance against the Air Force concerning the issues of close air support and tactical and strategic airlift.[83] It perceived that the Air Force was disinterested in providing the necessary close air support and airlift. Taylor told a conference of senior Army commanders in 1956: "We haven't had close effective tactical air support; we cannot expect to have it in the future. The high-performance Air Force planes are flying from us; they have left the battlefield."[84] Perhaps, as a result, the Army had a hidden agenda to develop its own "air force." Two important Army Chiefs of Staff of the fifties, Matthew Ridgway and Maxwell Taylor, reinforced this.

In a book published in 1956 after he retired, Ridgway was adamant that somebody must man these close support and assault aircraft. He said that "the Army, of course, has considered seeking to relieve the Air Force of its unwanted burden. Plans have been advanced whereby the Army would develop its own specialized assault aircraft, and recruit and train its own pilots to fly them, and to a slight degree this has been done." He warned that if the Air Force "continue to ignore our needs in this respect, we eventually will have to develop them ourselves."[85] Similarly, Taylor published a book in 1959 after he retired. In it he stated that the Army "should have its own organic tactical air support and tactical air lift."[86] Ridgway's and Taylor's statements implied that the Army had been trying to regain the tactical air support lost since the Air Force became a separate service. This notion of the Army trying to substitute the tactical air support provided by the Air Force has wide currency.[87]

It is curious that Ridgway and Taylor made such a claim after they retired. Official statements made while they were serving (including by Brucker) never openly vied for the Air Force roles and missions.[88] Nevertheless, these statements and ones made by serving Army officers who disagreed with, or who simply did not know, the official position, fueled Air Force suspicions.[89] Moreover, Army aviation developments in the fifties, plus a dispute during Exercise SAGE BRUSH,[90] reinforced the Air Force's perception of the Army's intention to usurp its functions. For example, Army transport helicopter developments made the Air Force suspect that the Army intended to usurp its air transport function for airborne operations, or eventually the entire airlift function. The Army also tested armed fixed wing airplanes and helicopters, and had a program to develop helicopter armaments. This made the Air Force suspect that the Army was attempting to encroach upon its close air support function. Moreover, the Army wanted to obtain a "higher performance" observation aircraft for deep penetration missions to acquire targets for its missiles. The Air Force construed this as an attempt at usurping the tactical reconnaissance function. These disputes intensified and climaxed in 1955–1956, when the Air Force had to relinquish its transport helicopter units.[91] It then prepared itself for an all-out Army effort to gain control of tactical air forces.[92] Charles Wilson, the Secretary of Defense from 28 January 1953 to 8 October 1957, had to intervene on 26 November 1956 to reduce the confrontation (see next chapter).

It is undeniable that interservice rivalry existed over the issue of Army aviation. Interservice rivalry as a general phenomenon increased appreciably after World War II as a result of "unifying" the three services under one department. During the early years the Department of Defense was bureaucratically weak in comparison with the individual services. It lacked strong hierarchical control over the service departments. This translated into lax fiscal and policy control from the Office of the Secretary of Defense and the Joint Chiefs of Staff. Apart from setting a budgetary ceiling and making decisions on strategic programs, the OSD allowed each service to spend its money according to its own funding priorities. This allowed the interservice rivalry to flourish.[93]

However, the 1958 Defense Reorganization Act put a stop to this practice. This act weakened the role of individual services in decision making while strengthening the OSD and the JCS in research and development policy making. It also strengthened the relationship between the JCS and the specified and unified commands. It strengthened the major commands as combat organizations, while making the services become more administrative organizations. After the Defense Reorganization Act of 1958 the level of interservice rivalry subsided. After 1961 the Kennedy-Johnson administration introduced even tighter fiscal control and policy control over the military services. Defense Secretary Robert McNamara shifted the balance of influence even further away from the service departments to his own office.[94] With decreased powers in research, development, and procurement, the individual services had less "slack" and bureaucratic strength with which to "do their own thing" and thus clash over issues like Army aviation.[95]

On balance, there is insufficient evidence for arguing that the Army's desire to build up its aviation capabilities was primarily to make up for Air Force neglect. There is instead a strong body of evidence to suggest that the need for its own aviation force came from other factors. These included genuine combat utility, the exigencies and lessons of the Korean war, and the political realities of the fifties and sixties. It was more likely that the Army was developing its aviation programs and air mobility for solving its own overriding problems on the battlefield.[96]

NOTES

1 (1) John Lewis Gaddis, *Strategies of Containment: A Critical Appraisal of Postwar American National Security Policy* (New York: Oxford University Press, 1982), pp. 147–53, 166–67. (2) Samuel P. Huntington, *The Common Defense: Strategic Programs in National Politics* (New York: Columbia University Press, 1961), pp. 64–113.

2 (1) Galvin, pp. 274–75. (2) A. J. Bacevich, *The Pentomic Era: The US Army Between Korea and Vietnam* (Washington DC: National Defense University Press, 1986), pp. 15–21, et passim.

3 Huntington, pp. 73–76.

4 (1) Bacevich, pp. 21 ff. (2) Huntington, p. 421.

5 Taylor said that "nuclear weapons were the going thing and, by including some in the division armament, the Army staked out its claim to a share in the nuclear arsenal." Maxwell D. Taylor, *Swords and Plowshares* (New York: W. W. Norton & Company, 1972), p. 171.

6 Huntington, p. 405.

7 Matthew Evangelista, *Innovation and the Arms Race: How the United States and the Soviet Union Develop New Military Technologies* (Ithaca, New York: Cornell University Press, 1988), pp. 149, 207.

8 DA, "Army Equipment Development Guide, Changes No. 1," p. 1.

9 House of Representatives, *Army Appropriations FY 54*, pp. 1473–74.

10 House of Representatives, *Army Appropriations FY 54*, p. 1473.

11 Theodore H. White, "An Interview with General Gavin . . . Tomorrow's Battlefield," *The Army Combat Forces Journal*, March 1955, p. 20.

12 Matthew B. Ridgway, *Soldiers: The Memoirs of Matthew B. Ridgway* (New York: Harper & Brothers, 1956), p. 42.

13 White, p. 23.

14 (1) Major Robert A. Doughty, US Army Command and General Staff College, Combat Studies Institute, Leavenworth Papers No. 1, *The Evolution of US Army Tactical Doctrine, 1946–76* (Fort Leavenworth, Kansas: GPO, August 1979), p. 16. (2) DoD, *Semiannual Reports, Jan.–June, 1954*, p. 78. (3) Richard P. Weinert, Historical Office, Office of the Chief of Staff, US Army Training and Doctrine Command, *A History of Army Aviation 1950–1962, Phase II: 1955–1962* (Fort Monroe, Virginia, November 1976), pp. 60–61. Hereafter all Weinert references refer to this work. (4) E. F. Fisher, Office of the Chief of Military History, *Relationships of the ROAD Concept to Moral Considerations in Strategic Planning* (Washington, DC: GPO, 28 October 1964), pp. 42–48.

15 (1) Colonel Wayne E. Downing, Commanding Officer, 30th Engineer Group (Topographical Survey), "Combat Mobility with Army Aircraft," *MR*, July 1956, p. 48. (2) G. J. McAllister, "Army Wants Air 'Cavalry' and 'Trucks,'" *AW*, 14 March 1955, pp. 90–91.

16 Ridgway, *Soldiers*, pp. 298–99.

17 Brigadier General Paul F. Yount, Acting Chief of Transportation, "Air Transportation and the Army," *NDTJ*, September–October 1953, pp. 63, 64.

18 White, p. 23.

19 (1) Weinert, p. 61. (2) DoD, *Semiannual Reports, Jan.–June, 1956*, p. 87. (3) Doughty, p. 16. (4) General Maxwell D. Taylor, "The Army Rewrites the Book," Address to the Pikes Peak Chapter, Association of the United States Army, Colorado Springs, Colorado, 29 May 1957.

20 (1) Wilber M. Brucker, Secretary of the Army, "Why the Army Needs Aviation," *AA*, 10 March 1958, p. 21. (2) Major General Hamilton H. Howze, Director of Army Aviation, Office of the Deputy Chief of Staff for Military Operations, "Army Aviation Looks Ahead," *AID*, July 1957, pp. 3–4. (3) Major General John P. Daley, Director of Special Weapons, Office, Chief of Research and Development, United States Army, "Mobility Is a Must," *Ordnance*,

March–April 1958, p. 797. (4) United States Army, The Infantry School, "Infantry Instructors' Conference Report," Fort Benning, Georgia, 4–9 June 1956, pp. 84–98. Hereafter cited as Infantry School, "1956 Instructors' Conference."

[21] (1) "The Chief Speaks on Organic Air," *ANAFJ*, 28 July 1956, p. 11. (2) General Maxwell D. Taylor, "The Veteran and the Army," Address Before the Annual Convention of the Military Order of the World Wars, Annapolis, Maryland, 19 October 1956. (3) "CofS Taylor Foresees New Army Planes; 'Ground Vehicles Without Footprints,'" *ANAFJ*, 1 June 1957, p. 1. (4) General Maxwell D. Taylor, Chief of Staff, United States Army, "The Army of the Future," *AID*, January 1957, pp. 15–16.

[22] House of Representatives, *Army Appropriations FY 58 Part 1*, p. 147.

[23] General Maxwell D. Taylor, "The Army's Need for Mobility," Address at the Twelfth Annual Convention of the National Defense Transportation Association, Shoreham Hotel, Washington, DC, 18 November 1957.

[24] (1) Major Lawrence L. Mowery and Captain James B. Hobson, "ROCID Air Mobility," *Infantry*, July 1957, pp. 40–43. (2) DoD, *Semiannual Reports, Jan.–June, 1957*, pp. 95–96.

[25] DoD, *Semiannual Reports, Jan.–June, 1956*, pp. 75–85.

[26] (1) General Maxwell D. Taylor, Chief of Staff, "'Safety Lies Forward'—Technologically and Tactically," *Army*, December 1956, pp. 20–24. (2) Lieutenant General C. D. Eddleman, Deputy Chief of Staff for Military Operations, "Men, Missiles and Atomics on the Futurarmy Battlefield," *Army*, December 1956, pp. 24–30. (3) Lieutenant General C. D. Eddleman, Deputy Chief of Staff for Military Operations, "Military Operations," *AID*, February 1957, pp. 17–23.

[27] (1) United States Army Infantry School, "Infantry Instructors' Conference Report," Fort Benning, Georgia, 17–21 June 1957, p. 23. Hereafter cited as Infantry School, "1957 Instructors' Conference." (2) Colonel Clinton E. Granger, Jr, United States Military Academy, "Concepts of Mobility," *MR*, February 1960, pp. 45–51.

[28] Yount, "Air Transportation," p. 62.

[29] General Matthew B. Ridgway, "An Army on Its Toes," *The Army Combat Forces Journal*, December 1954, p. 10.

[30] Lieutenant Colonel John C. H. Lee, Jr, Member, Joint Airborne Troop Board (with Sidney Shalett), "We Must Put Wings on the Infantry," *Saturday Evening Post*, 17 May 1952, p. 42.

[31] (1) Wilber M. Brucker, Secretary of the Army, "Look at the Army Whole," *Army*, December 1957, p. 26. (2) Wilber M. Brucker, Secretary of the Army, "The Army of the Future," *AID*, January 1957, p. 14.

[32] "Army Operations in Limited War," *AID*, June 1958, pp. 41–42.

[33] (1) G. J. McAllister, "Army Puts New Stress on Air Mobility," *AW*, 15 March 1954, pp. 65–70. (2) McAllister, "Army Wants Air 'Cavalry' and 'Trucks,'" pp. 89–90.

[34] Ridgway, "An Army on Its Toes," p. 10.

[35] Ridgway, *Soldiers*, p. 312.

36 (1) DoD, *Semiannual Reports, Jan.–June, 1955*, p. 88. (2) Department of the Army, Joint Airborne Troop Board, "Report of the Conference on Joint Airborne Matters," Fort Bragg, North Carolina, 17–20 November 1953, Appendix A1, p. 44; and covering letter. Hereafter cited as JATB, "1953 Conference." (3) Department of the Army, "Army Equipment Development Guide," Washington 25, DC, 3 May 1954, pp. 5, 97–99.

37 (1) Infantry School, "1956 Instructors' Conference," p. 124. (2) Infantry School, "1957 Instructors' Conference," p. 26.

38 General Maxwell D. Taylor, Address at Tenth Annual Convention, National Defense Transportation Association, Sheraton Plaza Hotel, Boston, Massachusetts, "The Army and Mobility," 13 October 1955.

39 General Maxwell D. Taylor, "The Changing Army," *The Army Combat Forces Journal*, October 1955, p. 10.

40 Huntington, pp. 352–53.

41 (1) Wilber M. Brucker, Secretary of the Army, "Look at the Army Whole," *Army*, December 1957, p. 26. (2) Wilber M. Brucker, Secretary of the Army, "How Far We Have Come!," *Army*, September 1960, p. 40. (3) Major General Earle G. Wheeler, Director of Plans, Office of the Deputy Chief of Staff for Military Operations, "Strategic Mobility," *AID*, January 1957, pp. 5–6.

42 Taylor, "The Army and Mobility."

43 General Lyman L. Lemnitzer, Chief of Staff, US Army, Address to the Seventh Annual Conference of Civilian Aides to the Secretary of the Army, Aberdeen Proving Ground, Maryland, "The Army in the 1960's," 16 September 1960.

44 Lieutenant General Arthur G. Trudeau, Chief of Research and Development, Department of the Army, "Toward a Mobile Army," *NDTJ*, November–December 1959, p. 37.

45 Galvin, pp. 274–75.

46 (1) Infantry School, "1957 Instructors' Conference," p. 43. (2) United States Army Aviation School, "Annual Army Aviation Instructors' Conference, 1960: Report of Proceedings," Fort Rucker, Alabama, 26–28 July 1960, pp. 14–15. Hereafter cited as Aviation School, "1960 Instructors' Conference." (3) Lieutenant General Arthur G. Trudeau, Chief of Research and Development, "Air Mobility Requirements," *Army*, June 1959, p. 58.

47 "Army's New Aviation Chief Deplored 'Starvation Basis' Support; B/G von Kann Wants No Army-AF 'Jurisdictional Dispute,'" *ANAFJ*, 13 June 1959, p. 18.

48 "Army Urged to Boost Use of 'Air Medium,'" *ANAFR*, 5 September 1959, p. 12.

49 Major General Ernest F. Easterbrook, Commanding General, US Army Aviation Center, "Army Aviation Looks Ahead," *USAAD*, December 1959, pp. 1–4.

50 (1) Tolson, *Airmobility*, p. 4. (2) Colonel George P. Seneff, Jr, Air Mobility Division, Office, Chief of Research and Development, "Mixture for Mobility," *Army*, March 1959, p. 49.

[51] (1) Robert Farrell, "Special Report from Algeria, Part I: French Meet Guerrillas with Helicopters," *AW*, 17 September 1956, pp. 28–31. (2) Robert Farrell, "Special Report from Algeria, Part II: Algerian Terrain Challenges Helicopters," *AW*, 24 September 1956, pp. 88–92. (3) Robert Farrell, "Special Report from Algeria, Part III: Maintenance Plagues Fine Army Base," *AW*, 1 October 1956, pp. 73–81.

[52] Lieutenant Colonel John W. Oswalt, Deputy Director of Combat Developments, US Army Aviation Center, "Why and How Army Aviation Arms for Battle," *Army*, May 1958, p. 42.

[53] (1) "Observers Report on French Army Helicopter Combat Experience in Algeria," *ANAFJ*, 25 May 1957, p. 11. (2) See also *ANAFJ*, 18 May 1957, p. 1.

[54] (1) Brigadier General Bogardus S. Cairns, Address to the Military Order of World Wars, Fort Benning, no title, 20 February 1958. (2) See also: Brigadier General Bogardus S. Cairns, Commanding General, United States Army Aviation Center, "Army Aviation = Battlefield Mobility," *AID*, August 1958, p. 10.

[55] Yount, "Air Transportation," pp. 63–64.

[56] White, p. 23.

[57] (1) Tolson, *Airmobility*, p. 4. (2) Seneff, pp. 47–49.

[58] Major General James M. Gavin, "Cavalry, and I Don't Mean Horses!" *Harper's Magazine*, April 1954, pp. 54–60.

[59] Gavin, "Cavalry, and I Don't Mean Horses!," pp. 54–60.

[60] (1) Captain James M. Rockwell, "Future," *USAAD*, August 1958, p. 8. (2) Lieutenant Colonel Edgar C. Wood, Army Aviation Division, Assistant Chief of Staff, G-3, Department of the Army, "The Army Helicopter, Past, Present and Future," *Journal of the American Helicopter Society*, January 1956, p. 87.

[61] (1) Brigadier General Carl I. Hutton, Commandant, Army Aviation School, "Imbalance Between Firepower and Mobility," *USAAD*, August 1955, pp. 3–4. (2) Brigadier General Carl I. Hutton, Commanding General, Army Aviation Center, "Joint Strategy," *USAAD*, July 1956, pp. 3–4.

[62] "Army Operations in Limited War," p. 44.

[63] (1) Major General Hamilton H. Howze et al., "The Last Three Years of Army Aviation," *USAAD*, March 1958, p. 58. (2) McAllister, "Army Wants Air 'Cavalry' and 'Trucks,'" pp. 90–91. (3) Colonel Stewart L. McKenney, G2, XVIII Airborne Corps, Fort Bragg, North Carolina, "SKYCAV Operations During Exercise SAGEBRUSH," *MR*, June 1956, pp. 11–12.

[64] Department of the Army, Office of the Deputy Chief of Staff for Military Operations, "Army Aviation Plan FY 56 – FY 60," Washington, DC, 5 March 1956, Annex 1, Appendix D, p. 1. Hereafter cited as DA, "1956 AAP."

[65] Lieutenant General John C. Oakes, Deputy Chief of Staff for Military Operations, "Facets of Mobility," *Army*, September 1960, p. 51.

[66] See as examples: (1) Brigadier General Clifton F. von Kann, Director of Army Aviation, Office of the Deputy Chief of Staff for Military Operations, "Mobility Now," *AID*, April 1960, pp. 32–33, 38, 40. (2) Brigadier General

Clifton F. von Kann, Director of Army Aviation, ODCSOPS, "The Real Goal of Army Aviation," *USAAD*, March 1960, pp. 35–36.

[67] (1) Major Harold D. Pinney, US Army Infantry School, "The Soldier on the Battlefield," *Army*, September 1960, pp. 64–65. (2) Colonel G. Harry Huppert, Combat Developments Office, US Army Infantry School, "A Nonrated Infantryman Looks at Army Aviation," *USAAD*, March 1961, p. 14.

[68] Captain Robert B. Asprey, "Wintershield 2: War in a Nuclear Climate," *Army*, May 1961, pp. 40–46.

[69] (1) Weinert, pp. 70–71. (2) Fisher, pp. 57–58. (3) Bacevich, pp. 134–45.

[70] (1) Richard A. Aliano, *American Defense Policy from Eisenhower to Kennedy: The Politics of Changing Military Requirements* (Athens: Ohio University Press, n.d. [c.1975]), pp. 243–45. (2) "Military Will Command Major Attention in Congress; Both Parties Pledge Action," *ANAFJ*, 6 August 1960, pp. 1, 34.

[71] Gaddis, pp. 174–75.

[72] "TAC Commander Calls Army Aviation 'Second Class Force'; General Everest Reports Talk of New Specified Command," *ANAFJ*, 1 October 1960, p. 13.

[73] "Lemnitzer to Chairman, JCS," *Army*, September 1960, p. 14.

[74] See "Army Operations in Limited War," p. 42.

[75] General Lyman L. Lemnitzer, Chief of Staff, United States Army, Address at the Annual Banquet, 89th Annual Meeting, National Rifle Association, Sheraton-Park Hotel, Washington, DC, "The Rifleman and Modern War," 23 March 1960.

[76] Lemnitzer, "The Army in the 1960's."

[77] Lemnitzer, "The Army in the 1960's."

[78] (1) "B/G von Kann Sees Army Aviation as Key to Survival; Discusses Obstacles in Path," *ANAFJ*, 11 July 1959, p. 20. (2) See also: "Aviation that will Serve the Soldier on Battlefields Nuclear or Non-Nuclear," *Army*, January 1960, p. 11.

[79] Brigadier General Clifton F. von Kann, "Army Aviation," *Army*, November 1960, pp. 43, 47.

[80] (1) Fisher, pp. 58–63. (2) Weinert, p. 71. (3) Doughty, pp. 19–20.

[81] (1) Colonel Metcalf, Presentation to the Combat Developments Logistical Conference, HQ USCONARC, "Broad Briefing—MOMAR Study," 29 April 1959, pp. 1–2, 13–15; in Headquarters, United States Continental Army Command, Fort Monroe, Virginia, Memorandum for Distribution, "MOMAR Working Papers," 22 May 1959. (2) United States Army Aviation School, "Assistant Commandants' Conference on Developmental Objectives for Army Aviation 1959–1970," Fort Rucker, Alabama, 29 April 1959, pp. 1–5. Hereafter cited as Aviation School, "Conference on DOAA."

[82] (1) Fisher, p. 63. (2) Weinert, p. 72. (3) Doughty, pp. 20–21.

[83] (1) Bergerson, pp. 16–17. (2) Michael H. Armacost, *The Politics of Weapons Innovation: The Thor-Jupiter Controversy* (New York: Columbia University Press, 1969), pp. 39–44.

[84] Cited in Bacevich, p. 88.

[85] Ridgway, *Soldiers*, pp. 314–15.

86 Maxwell D. Taylor, *The Uncertain Trumpet* (New York: Harper & Brothers, 1959), p. 169.

87 (1) Morton Halperin, *Bureaucratic Politics and Foreign Policy* (Washington, DC: Brookings Institution, 1974), pp. 43–46. (2) Doughty, p. 27. (3) Andrew F. Krepinevich, Jr, *The Army and Vietnam* (Baltimore: Johns Hopkins University Press, 1986), pp. 114–15.

88 (1) "No AAC Revival Sought—Brucker," *AFT*, 26 November 1955, p. 2. (2) Claude Witze, "Army Budget Stresses Planes, Missiles," *AW*, 16 January 1956, p. 31. (3) "Army Seeks No New 'Air Corps,' Says Gen. Gavin; Stresses Mobility Need," *ANAFJ*, 28 January 1956, p. 35. (4) Evert Clark, "Army Outlines Aviation Program, Sees No Conflict with USAF," *AW*, 4 June 1956, p. 34. (5) "Army Aviation Seeks Faster, More Powerful Planes; Asks Added R&D Funds," *ANAFJ*, 14 July 1956, p. 11.

89 Goodhand, "Army Aviation Policy," pp. 1–2.

90 (1) United States Air Force, "History of the Tactical Air Command, 1 January 1955–30 June 1955, Volume III: Tactical Airlift: Requirements and Organizations," Langley Air Force Base, Virginia, pp. 28–35, and documents no. 6 to 12. (2) Weinert, p. 122. (3) See also *AW*, *ANAFJ*, and other press reports at the time (November–December 1955).

91 (1) United States Air Force, "History of 314th Troop Carrier Wing (Medium), Eighteenth Air Force for 1 January–30 June 1956, Vol. 1," Sewart Air Force Base, Tennessee, n.d., pp. 2, 17–18, 22–25. (2) Senate, *DoD Appropriations FY 57*, p. 918. (3) House of Representatives, *Air Force Appropriations FY 57*, pp. 854–55.

92 (1) The Evaluation Staff, Air War College, Air University, "Tactical Air Forces and National Security," Maxwell Air Force Base, Alabama, April 1955. (2) Headquarters Tactical Air Command, Staff Study, "A Solution to the Problem of Resolving Air Force–Army Doctrinal Differences Pertaining to Command, Control, and Employment of Air Forces, and to Prevent Infringement Upon and Unnecessary Duplication of Air Force Functions by Present and Proposed Army Organic Aviation," n.d. [c.1956].

93 (1) Alain C. Enthoven and K. Wayne Smith, *How Much is Enough? Shaping the Defense Program 1961–1969* (New York: Harper & Row, 1971), pp. 11–21. (2) Robert S. McNamara, *The Essence of Security: Reflections in Office* (London: Hodder & Stoughton, 1968), pp. 89–90. (3) Armacost, pp. 72–81, 281. (3) Huntington, pp. 151, 190, 378–81.

94 (1) Armacost, pp. 236–37, 278, 284–85, 288. (2) Huntington, p. 435.

95 (1) William W. Kaufmann, *The McNamara Strategy* (New York: Harper & Row, 1964), pp. 202–3, 243–45, 292–93. (2) Halperin, pp. 51–52.

96 Robert A. Olson, "Air Mobility for the Army: A Case Study of Policy Evolution in Relation to Changing Strategy," *Military Affairs*, Winter 1964–1965, pp. 169–70.

SOURCES FOR TABLES

3-1. Department of the Army, Office of the Deputy Chief of Staff for Military Operations, "Army Aviation Guidelines for the Development of Doctrine and Organization Through FY 1963," Washington, DC, 19 September 1958, Annex 1, Appendix G, p. 1. Hereafter cited as DA, "1958 AAG."

3-2. Mowery and Hobson, pp. 43–44.

4

HIGH-LEVEL SUPPORT FOR AIR MOBILITY

During the Eisenhower period the Army high command maintained a high-level investment in air mobility. It tried to retain the plan for 12 transport helicopter battalions. The post-Korea reduction in the Army force structure, however, affected the air mobility program. Since there were fewer ground units to support, the Army adjusted its air mobility plans downwards. Nevertheless, the revised aviation plan was still sizable. This impressive determination to obtain a large aviation force structure was possible even though there was a reduction in the Army's overall budget share. Contrary to popular presumption, the Eisenhower administration's fiscal policies provided sufficient funding for the Army aviation program. Apart from matériel issues, the Army also made significant progress in doctrinal terms. The Army high command gradually transformed air mobility into a well-rounded concept, encompassing both tactical and logistical aspects. It also became increasingly mainstream, with the publication of field manuals and becoming part of the service school curriculum. Even strong Air Force opposition did not deflect the Army's progress with air mobility. This was because the Department of Defense took a favorable view toward Army aviation. New "restrictions" for this period proved not to be an obstacle for air mobility development.

THE SUFFICIENCY OF FUNDING AND PROGRAM SUPPORT

The Eisenhower Administration's Fiscal Policies

The central tenet of the New Look policy was to reduce military expenditures to a long-term, sustainable level. Conventional war-fighting capabilities received less funding, and the Army received the smallest budget of the three services. In relative terms it received about half the size of the Air Force budget, or only 23 percent of the total defense expenditure (see Table 4-1).[1] Hence the Army had to cut its manpower continuously. Many later saw this period of Army history as one of unjustified austerity and strategic neglect. They felt the Army lacked capital investment in weapons, and this affected air mobility developments as well. Statements to Congress such as the following from Taylor in 1958 reinforce this impression:

> I have often had occasion to say if we had no limitations on Army aviation the size of our budget would never let us go very far in Army aviation. . . . We find in the budgets we have been dealing with the last few years, about $125 million or $130 million is the most we could put in the entire program.[2]

However, the $125 million per year was adequate to sustain a decade of growth. It was still quite a sizable budget, being slightly more than the Korean war average (see Table 4-2). Even the total Army procurement and production expenditures were 2.5 times the

Table 4-1
Defense Expenditures, FY 1948–1961 (in Billions of Dollars)

Fiscal Year	Army	Navy	Air Force	OSD	Total
1948	5.7	4.3	1.1	0.0	11.1
1949	5.6	4.4	1.8	0.0	11.9
1950	–	–	–	–	11.9
1951	7.5	5.6	6.3	0.4	19.8
1952	15.7	10.2	12.7	0.4	39.0
1953	16.3	11.9	15.1	0.4	43.7
1954	12.9	11.3	15.7	0.4	40.3
1955	8.9	9.7	16.4	0.5	35.5
1956	8.7	9.7	16.8	0.6	35.8
1957	9.0	10.4	18.4	0.6	38.4
1958	9.1	10.9	18.4	0.7	39.1
1959	9.5	11.7	19.1	0.9	41.2
1960	9.4	11.6	19.1	1.1	41.2
1961	9.6	11.8	18.9	1.2	41.5

Table 4-2
Army Aviation Budget, FY 1954–1960, Obligations Incurred, Procurement of Equipment and Missiles, Army (PEMA)

Fiscal Year	Army Aircraft	Major Equipment	PEMA Total	% Major Equipment	% PEMA Total
1954	N/A	$62,095,442	$1,120,779,819	–	–
1955	$130,614,077	813,407,186	2,322,681,649	16.1	5.6
1956	121,197,335	N/A	1,893,529,000	–	6.4
1957	122,347,811	745,661,460	2,072,886,018	16.4	5.9
1958	N/A	532,289,434	1,313,282,000	–	–
1959	54,008,221	391,107,497	1,096,954,530	13.8	4.9
1960	[152,000,000]	770,645,400	1,387,584,000	19.7	11.0

Korean war average (see Table 4-3). This meant that the end of the Korean war did not spell a decline in Army aircraft procurement. The fleet of Army aircraft increased 70 percent over this period (see Table 4-4). Even the aircraft research and development budget received adequate funding to cover important projects (see Table 4-5).

What the administration policies caused was a reduction in the aviation force structure goals established during the Korean war. Fewer personnel and fewer ground units required less aircraft and aviation units for support. Despite the reduction in aircraft inventory goals, the number of aircraft per division slice was steady over the 1956–1958 period (see Table 4-6). Moreover, the reduction in inventory goals did not matter since the actual inventory before 1958

Table 4-3
Army Procurement and Production Budget, Funds Available and Expenditures, FY 1951–1965 (in Millions of Dollars)

Fiscal Year	Carryover	(Unobligated)	New Appropriations	Expenditures
1951		–	–	–
1952	10,363	(10,363)	8,726	580
1953	9,781	(1,374)	2,736	589
1954	11,927	(1,713)	3,225	4,399
1955	10,753	(5,508)	0	2,628
1956	8,328	(4,874)	0	1,584
1957	5,636	(2,525)	0	1,450
1958	4,086	(1,596)	20	1,547
1959	2,917	(442)	1,712	1,627
1960	2,635	(504)	1,407	1,604
1961	2,325	(649)	1,495	1,526
1962	2,440	(501)	2,532	1,815
1963	3,198	(873)	2,519	2,370
1964	2,799	(973)	2,906	2,315
1965	3,391	(1,758)	1,783	1,764

Table 4-4
Army Aircraft Inventory, FY 1953–1960

Fiscal Year	Fixed Wing	Rotary Wing	Total	% Increase	No. Increase
1953	N/A	N/A	3,200	–	–
1954	N/A	N/A	3,633	13.5	433
1955	2,351	1,188	3,539	-2.6	-94
1956	2,117	1,456	3,573	1.0	34
1957	2,546	1,901	4,447	24.5	874
1958	2,834	2,193	5,027	13.0	580
1959	2,842	2,357	5,199	3.4	172
1960	2,860	2,633	5,493	5.7	294

was well below those goals. By FY 1960 the actual number of aircraft exceeded the 1958 goal. The 5,493 aircraft supporting 14 divisions meant there were 392 aircraft per division slice.

The goals for transport helicopter units also declined. In addition, the 1957 and 1958 guidelines had a noticeable shift towards light transport helicopters and fixed wing transports. This was a reflection of the heavy transport helicopter's immature technology, and the inadequate production capacity for medium transport helicopters, rather than a cost-cutting measure (see next chapter).[3] The goal for transport aircraft lift per division slice therefore declined over the 1956–1958 period. However, increases in the number of transport-type aircraft within the division partially compensated for this. Moreover, by 1960 there was a reversal of the fall in the number of aviation companies (see below).

Plans notwithstanding, Army aviation made good progress in increasing the proportion of load-carrying aircraft in its inventory (see Table 4-7). In July 1955 only 16.5 percent of the total aircraft fleet were utility airplanes, cargo airplanes, utility helicopters, or

Table 4-5
Army Aviation Budget, FY 1953–1959, Obligations Incurred, Research and Development, Army

Fiscal Year	Aircraft & Related Equipment	Total Direct Obligation	% of Obligation
1953	N/A	?	–
1954	$20,459,000	$365,517,395	5.6
1955	7,700,657	366,324,517	2.1
1956	15,768,843	403,548,129	3.9
1957	10,705,882	427,397,176	2.5
1958	59,954,381	801,282,465	7.5
1959	54,484,996	1,021,524,232	5.3

Table 4-6
Key Goals of Army Aviation Plans, 1954–1958

	1954	1956	1957	1958
Total division force structure	20	19	16	15
FY 1960 inventory goal	8,500*	6,161	5,333	5,297
Number of transport battalions^	12 (3)	12 (3)	7 (5)	6 (5)
Number of heavy companies†	–	9 (21)	0	0
Number of medium companies†	–	9 (21)	7 (16)	6 (16)
Number of light companies†	–	18 (21)	21 (20)	19 (20)
Number of fixed wing transport companies†	0	3 (21)	7 (16)	6 (16)
Total number of transport companies	36	39	35	31
Aircraft per division slice	425	324	333	353
Tons per lift per division slice		114	70	66

* By year 1959
^ Parenthesis denotes companies per battalion
† Parenthesis denotes aircraft per company

cargo helicopters. By October 1956 the proportion increased to 27.9 percent. By June 1958 this tilted even more towards transport-type aircraft—33.3 percent. By June 1959 it was 34.8 percent. This increase was all the more impressive considering the higher cost of larger aircraft. In 1960 Major General Frank Besson, Jr, the Army Chief of Transportation, said: "In order to get the payload we need, our aircraft have necessarily become bigger, and heavier. Unfortunately, this also makes them more expensive to buy, maintain and operate. This trend is emphasized by the fact that while, under present plans we expect the Army aircraft population to begin to level off this year, we find that the dollar value continues to rise."[4] A sustained procurement program of high-cost aircraft would not have been possible without the strong support from the Army high command.

It is therefore difficult to claim that the Army or Army aviation suffered from a severe funding squeeze. Indeed, the Army kept many of its other key projects relatively well funded. For example, the Army spent an inordinate amount on ballistic missiles, although this was

Table 4-7
Army Aircraft Assets On Hand, 1955–1959

	Airplanes				Helicopters		
Date	Obsn	Utility	Cmnd	Cargo	Recon	Utility	Cargo
1 July 1955	1,818	360	97	6	1,108	118	115
1 October 1956	1,673	541	110	66	1,132	172	348
30 June 1958	1,973	603	171	111	1,251	256	728
30 June 1959	1,967	629	186	152	1,373	305	796

mostly for political and prestige reasons. In addition, the Army developed and introduced thousands of armored personnel carriers during this period. This was another costly item that previous Army procurement budgets did not support to this extent. The Army also initiated development of new self-propelled artillery and new tanks, while rectifying those rushed into production during the Korean war without proper testing.

An important mitigating circumstance was the Department of Defense's lax fiscal and policy control over the Army. Able to spend the available funds as it saw fit, the Army was able to realign funding priority towards capital equipment for air mobility, given the budgetary ceiling. Undoubtedly, Ridgway, Taylor, Lemnitzer, and others in the Army high command willingly invested in Army aviation, second only to missiles.[5] This high-level attention accounted for the great progress in air mobility.

Support from Ridgway

After taking over from General Collins, Ridgway continued to support the existing Army aviation programs, especially the 12-battalion transport helicopter program.[6] Ridgway also instigated a review of Army aviation. This included a five-year Army Aviation Plan that called for 8,486 aircraft by 1959, a substantial increase from the 1954 inventory of 3,516.[7] His review on Army operations under nuclear conditions also demonstrated his pro–air mobility stance. His experience of Korea was no doubt the formative influence. The other influential officer in the Army high command, General Gavin, headed the review on nuclear operations. He also wrote the seminal article "Cavalry, and I Don't Mean Horses!" Indeed, Gavin envisaged that 20,000 transport aircraft might not be too many for a "cellular-type" Army.[8] Moreover, the Secretary of the Army, Robert Stevens (4 February 1953 to 21 July 1955), was also an ardent supporter of air mobility.[9] This also applied to the Assistant Secretary of the Army, Frank Higgins.[10] The Semiannual Report of the Secretary of the Army, for 1 January to 30 June 1955, emphasized the great importance of the continued build-up of Army helicopter companies. While this was to provide greater tactical mobility for only small units, personnel, and supplies within the combat zone, these companies "are the backbone of the Army's third-dimensional mobility."[11]

Besides obtaining more aircraft, the Army high command also tried to strengthen Army aviation bureaucratically. On 24 April 1954 the

Department of the Army established an aviation staff section within its central planning staff. This Army Aviation Branch came under the Organization and Training Division, Assistant Chief of Staff, G-3 (ACS, G-3). Then on 11 January 1955 the Army Policy Council approved several recommendations as part of Ridgway's review of Army aviation. First, on 1 February 1955 the Army Aviation Branch became a separate Army Aviation Division on a par with the Organization and Training Division and two others (Operations and Plans). The Department of the Army appointed Brigadier General Hamilton Howze as the first Chief on 28 February. He was not an Army aviator, and was not involved in aviation matters before this appointment. He was an ex-cavalry officer, graduate of West Point, and had served in cavalry, intelligence, and armor positions. His assignment before moving to the Pentagon was deputy chief of staff for plans, Seventh Army in Europe.[12]

The elevation of the staff section was due to the growth of Army aviation activities. This had arrived at the point where the Army had to take steps to ensure proper direction and control. The Army realized that Army aviation needed an effective general officer to defend it properly and to coordinate the disparate aviation structure within the Army. Later, on 3 January 1956, the Department of the Army again expanded the Army Aviation Division into a Directorate of Army Aviation, with Major General Howze continuing as its Director. This was the result of an Army reorganization that amalgamated the office of the ACS, G-3 with the Office, Deputy Chief of Staff for Plans. The new organization became the Office, Deputy Chief of Staff for Military Operations (ODCSOPS). Howze reported to Lieutenant General Clyde Eddleman, the Deputy Chief of Staff and a staunch supporter of Army aviation.[13] Eddleman played a critical part in the original Materiel Requirements Review Panel while he was the ACS, G-3 in 1952.[14]

The Army Policy Council also strengthened other aviation bureaucracies. On 1 February 1955 the Army established a strong aviation focal point at Camp Rucker (later Fort Rucker), the Army Aviation Center. Among other agencies, the Army sited the Army Aviation School and Army Aviation Board there. The CONARC Board Number 6 (or the Army Aviation Board) had the authority to test and experiment with anything related to aviation.[15] The Aviation School, apart from training aviators, was also responsible for developing and preparing for dissemination aviation doctrine, technique, and tactics.[16] Brigadier General Carl Hutton became the Commanding General of the Center, as well as Commandant of the School. He was a proponent of radical air mobility and initiated the development of armed helicopters and sky cavalry units. Under Hutton, the Center became a hotbed of radical ideas.

Elsewhere, the Department of the Army, G-1, received centralized control over aviation personnel. CONARC, too, and its subordinate service schools received significant assignments of aviation responsibilities, such as training, doctrine, and equipment (see below).[17] It is important to point out that Ridgway authorized these moves deliberately to strengthen the bureaucratic weight of Army aviation.[18]

By May 1955 General Howze said he got the impression of a strong and increasing support for Army aviation. He had no doubt that Army aviation was on an "ascending plane."[19]

Support from Taylor

Taylor not only supported current ongoing plans, he also approved another increase in Army aviation. The 1956 Army Aviation Plan increased the intended Army aircraft inventory objective for FY 1960 from 7,040 to 9,439.[20] Although the Army did not publicly reveal the details of this plan, on 12 February 1956 Taylor spoke of its significance on a national television program.[21] This officially approved plan stated six new functional guidelines for Army aviation. They were: observation; airlift for troop movement in the combat zone; rapid movement of critical supplies; air mobility for land reconnaissance; command, liaison, and communications; and battlefield casualty evacuation.[22] This was a significant departure from the functions defined in the 1952 Pace-Finletter agreement. Furthermore, it was more elaborate and specific than those outlined in the 1952 and 1954 Army Equipment Development Guides. It seems that Taylor was as radical as air mobility developments in the fifties would allow. He said in a speech:

> In developing Army tactical mobility, our progress is good but we are not yet satisfied. We are finding many tactical and logistical uses for our helicopters which lend flexibility to military operations. We are increasing the number both of helicopters and of fixed wing aircraft in the Army. Through the versatility of these vehicles, we are finding new ways to meet the requirements of supply and movement on the atomic battlefield.[23]

Moreover, during a congressional hearing on the 1959 Army appropriations, Representative Mahon asked General Taylor: "Do you think you are going too far in Army aviation?" He replied "No, sir, I do not. I should like to go further."[24] The only limit was budgetary, and the need to maintain personnel levels and fund firepower projects.[25] It seems that Taylor, like Ridgway, supported air mobility because he had been exposed to transport helicopters during Korea.[26] Another prominent and active supporter of Army aviation was Secretary of the Army Wilber Brucker (21 July 1955 to 20 January 1961).[27] In addition, the Assistant Secretary of the Army, George Roderick, was also an avid exponent of Army aviation. Roderick, Brucker, Taylor, Eddleman, and Gavin were among the officials who made decisions on the 1956 Army Aviation Plan.[28] The Secretary of the Army's 1956 Semiannual Report, reflecting the view of the entire Army leadership, declared: "Tactical victory will belong to the army with the superior mobility on a rolling battlefield; in the sense defined by future wars, aviation is a most important form of mobility."[29]

Apart from satisfying battlefield needs, another attraction of Army aviation was its "glamour value." The post-Korea fifties was a period of high profile scientific and technical exploitation. The Air Force was particularly good at using this to obtain more funds.[30] Once, Defense Secretary Wilson even sent the Army's budget back for being unexciting and asking for unglamorous weapons and equipment such as rifles, machine guns, trucks, and unsophisticated aircraft. Taylor said that Wilson wanted the Army to "substitute requests for newfangled items with public appeal instead of the prosaic accoutrements of the foot soldier."[31]

Support from Lemnitzer

General Lemnitzer also had a favorable disposition towards Army aviation. He was an early supporter of convertiplane development.[32] As the Deputy Chief of Staff for Plans and Research in the early fifties, he had exposure to Army aviation's matériel needs. As Chief of Staff, he initiated a major review of Army aircraft research, development, and procurement policy. Lieutenant General Gordon Rogers chaired this Army Aircraft Requirements Review Board, otherwise known as the Rogers Board. Part of the review process was to allow Army air mobility authorities to meet industry representatives. This, according to press reports, was to discuss "new and better ideas in aviation design concepts to meet the demands of limited war."[33] Lemnitzer approved the Rogers Board recommendations on 19 March 1960, which called for a ten-year Army Aircraft Development Plan.

The Rogers Board concluded that an optimum ten-year program would lead to an aircraft inventory increase to 9,500 and an increase to almost 15,000 aviators. However, this program was beyond the overall Army budget unless it underwent a substantial increase. Therefore the board reduced its objective to more realistic levels while at the same time still moving towards a significant degree of air mobility. The plan included a build-up of aviation spending to an annual plateau of $250 million, with three objectives in mind. The first was to provide a relatively high degree of air mobility for Army operations. The second objective was to provide a base upon which the Army could greatly increase its capabilities for air mobile operations after 1970. The third objective was to provide the field army with essential observation, target acquisition, combat surveillance, and reconnaissance capabilities. Relying on status quo assumptions of personnel, funds and political atmosphere, this program planned an aviation funding increase over the next four years until 1965. The Army would then reach about twice the 1960 annual expenditure.[34]

While the Rogers Board was still deliberating, General Lemnitzer decided to launch a build-up of 42 aviation companies to increase the mobility of ground troops. This reversed the cutbacks made in 1956, 1957, and 1958. Lemnitzer envisaged that, by the end of fiscal year 1965, the build-up to 42 companies would be complete. He disclosed this plan during congressional testimony in mid-February 1960.[35] Apart from Lemnitzer, other ardent supporters of air mobility included Lieutenant General Arthur Trudeau. He took over from Gavin as the Army Chief of Research and Development, and similarly advocated more and better air mobility.[36]

Congressional Support and the Aviation Lobby Group

Support for Army aviation came not only from the top Army hierarchy. Key figures in Congress were also very sympathetic. They included Senator Stuart Symington, Chairman of the Airpower Subcommittee, and Senator Henry Jackson, member of the same committee.[37] Army aviation's most ardent and long-time supporter in the House of Representatives was Daniel Flood, a member of the defense appropriation subcommittee.[38] Representative Robert Sikes was also a supporter. He charged that the Air Force was not capable of meeting the Army's requirement for tactical airlift, and asked what the Army was doing on its own to fill the gap.[39] Both Sikes and Flood even went so far as to encourage the Army to put up a stronger fight to reduce or eliminate the Department of Defense restrictions. Even Representative George Mahon, chairman of the House Military Appropriations Subcommittee, favored changing the restrictions.[40] Furthermore, the Senate Preparedness Subcommittee issued a report in 1960 detailing the budgetary support needed by the Army to accelerate the modernization of its aircraft programs.[41] One cannot emphasize enough the importance of congressional support. Congress could turn the political heat on the administration, or sometimes give more money to the services than the administration asked for. (The administration, however, could impound such funds and refuse to spend them.)

In early 1957 a small group of senior Army aviation officers in the active Army and Reserves, and members of industry, formed the Army Aviation Association of America (AAAA or Quad-A). These aviation-related interest groups created this nongovernmental organization to act as an external guardian of Army aviation interests. Membership of AAAA included not only serving members of Army aviation, it also included industrialists involved with Army aviation, and retired officers on their second career in military-related business. Although it claimed to be modeled on several professional-technical societies then in existence, in reality it was a quasi-lobbyist interest group. Its function was similar to that of, though narrower than, the Association of the United States Army (AUSA). Moreover, it was to foster camaraderie between aviators from different branches. The periodic national and local chapter meetings had a socializing effect similar to that of a professional body. It provided a forum for promoting and disseminating new ideas. The potency of this group became apparent within two years of its establishment. Apart from top Army and Army aviation officials, the press reported considerable congressional interest and involvement in this group at its first annual national meeting and banquet in June 1959.[42]

PROGRESS IN DOCTRINE

As the Army became increasingly aware of air mobility's importance in modern warfare, it became increasingly bold in its tactical usage. By the end of the decade, air mobility had achieved a degree of acceptance within mainstream Army doctrine. This was evident from the air mobility doctrine published and the curricula taught at service schools.

Increasing Awareness of Tactical Potential Under Ridgway

In January 1955 Major Robert Winkler, an Infantry officer with flying experience in Korea, commented: "The Transportation Corps, to which the transportation helicopter units belong, seems to consider the helicopter a substitute for the truck and emphasizes logistical support missions at the expense of tactical missions in the forward combat areas."[43] A statement by Colonel William Bunker, the Chief of Air Transport and Service, Transportation Corps, referring to the cargo helicopter as a flying truck reinforced this early bias.[44] Moreover, a 1952 manual from the Transportation School said:

> The employment of helicopter transportation will vary with the operational decisions of the commanders according to the situations existing at the time of employment. Helicopter units may be logistically employed in supply and evacuation missions or as part of a coordinated tactical operation with other air and ground forces. The advantages of employing helicopter units in tactical operations must be carefully weighed against the desirability of logistical use; it must be remembered that the cargo helicopter is relatively vulnerable, and estimated loss in achieving a tactical gain must be measured in terms of the possible loss of an invaluable logistical tool.[45]

The Transportation Corps was perhaps a little conservative when it came to the tactical applications of transport helicopters. However, action by the rest of the Army and the high command soon put paid to this. For example, in 1954 the Infantry School successfully challenged the term "cargo helicopter" (originally devised by the Transportation Corps in 1952) because of its logistical implications. Instead it suggested using the term "transport helicopter" to emphasize that the helicopter unit was a combat unit and the helicopter was a tactical vehicle.[46] The Army eventually adopted this term.

Moreover, on 9 July 1954 General Gavin urged the Chief of Army Field Forces to relegate helicopter logistical support to secondary importance. In the same letter, Gavin asked the combat arms to find applications for helicopter transport to meet their doctrinal needs. Army historian Richard Weinert said: "From this time onward, the Department of the Army became increasingly interested in the tactical applications of the helicopter rather than considering it as just another means of logistical transport."[47] On the same day, the Department of the Army requested OCAFF to prepare training literature for the employment of helicopter companies as tactical combat units. On 29 March 1955 the Department of the Army published Training Circular 1-7, *Employment of Army Transport Aviation*, prepared by CONARC. It replaced Training Circular 19 of 1950, *Transport Helicopter Company (Army)*. The new circular stressed the tactical function as primary, with logistical support or transport movement as an additional function. Army transport aviation was to move Army combat units operationally by air. This could involve airborne operations and air-landed operations.[48]

To support the new doctrine in the circular, in May 1955 CONARC initiated a long-range plan to prepare field manuals by various Army service schools. The Infantry School was to prepare, coordinate, and submit to CONARC a draft of a new transport aviation manual concerning combat operations. The Command and General Staff College was to change the Army's keystone manual, FM 100-5 *Operations*, to incorporate the new concept. It was also to produce a new transport aviation manual concerning logistical support operations. The Army Aviation School was responsible for a new manual covering the organization and operation of Army aviation transport units. The Chief of Transportation was responsible for a new manual covering the organization and operation of maintenance and supply units in support of Army aviation. By June 1955 the Infantry School completed and circulated the new draft manual, FM 57-35, *Army Transport Aviation Combat Operations*. This was a significant milestone as the difficulty of executing an air mobile operation without a proper tactical manual was apparent during an exercise in May 1954.[49]

By this time tactical air mobility also became a more common part of Army exercises. General Gavin inspired and urged the inclusion of a sky cavalry experiment in the 1955 Exercise SAGE BRUSH.[50] During Exercise BLUE BOLT (February 1955) conducted at Fort Hood, the 1st Armored Division made use of seven H-19 helicopters to perform tactical operations. The division inserted small detachments behind "Aggressor" lines to observe and report enemy movements. It also landed a small raiding party behind enemy lines and attacked a command post. The H-19 helicopters then performed aerial resupply of those observer detachments behind enemy lines.[51]

The 1955 Semiannual Report of the Secretary of the Army summed up the emphasis put on air mobility under Ridgway: "Particular attention was given to an effort to afford our combat forces a greater third-dimensional, vertical mobility which would permit envelopment of enemy forces from the air."[52] Later, several prominent Transportation Corps officers involved in the Army aviation program were also less conservative toward the idea of tactical air mobility.[53]

Balancing Logistical and Tactical Air Mobility Under Taylor

Under Taylor, doctrinal development of air mobility was more even, encompassing both tactical and logistical aspects. The main reason for the equal logistical and tactical emphasis stemmed from the need for divisions to be able to fight in dispersed islands of resistance. Subunits must be self-contained and semi-independent. Taylor said that "on the future battlefield with its great dispersion of units, Army aircraft will increase the effectiveness of combat operations and assist in overcoming increasingly complex supply operations."[54] Major General Howze added: "Resupply by Army aircraft will probably be routinely necessary for the most forward battle groups. Otherwise, supply would be very hazardous, because it will be impossible to prevent the infiltration into this area of enemy small parties which could easily disrupt the movement of our truck columns."[55]

Apart from having more transport aircraft, other factors made the aerial logistical support of highly dispersed units possible. First, there was an increase in the number of divisional aircraft and in their cargo-carrying capacity while there was a reduction in divisional strength and unnecessary heavy equipment. Moreover, the reorganization created nondivisional aviation companies at the corps and army levels, as well as within the armored cavalry regiments. This increased the number of aircraft in a nominal field army from over 900 to 1,747.[56] Furthermore, by 1957 the Army augmented this load-carrying capacity by the addition of bomb shackles on L-19 and L-20 airplanes. These shackles were able to carry external cargo rigged for free drop or parachute drop. The L-19 and L-20 both had two shackles, each with a 500-pound capacity. The Army demonstrated that using these airplanes for cargo-dropping missions were practical.[57] This was a significant increase in cargo-carrying capacity considering the number of L-19 and L-20 airplanes per division. Thus the amount of lift available in the Pentomic force structure was certainly significant.

The Army also tested the concept of air lines of communication and aerial resupply during Exercise STRONG ARM in May 1958 and found it a feasible proposition. This exercise involved an independent combat command in an exploitation operation. As a result, with

CONARC approval the Armor School was to develop doctrine for this type of operations.[58] Other tests also support the conclusion that Army transport aviation could provide aerial resupply to ground units.[59] For example, the Army's topographical survey operations in Alaska relied completely on Army aircraft for support because of the rudimentary ground infrastructure. The analogy to the expected future combat conditions was apparent to the Army.[60] Air logistics was therefore a requirement in theaters where ground infrastructure was primitive.[61]

Talk of transport helicopters replacing trucks wholesale, however, declined after 1955. Perhaps this was due to the realization that the cost of using helicopters was exorbitant.[62] Thus the Army never seriously considered the possibility of replacing huge supply dumps with transport aircraft. One of the six functions of Army aviation outlined in the 1956 Army Aviation Plan was "rapid movement of critical supplies." Presumably "routine" supplies did not require such exceptionally rapid movement. In 1957, Major General Yount said: "We are often asked how far we are going to go in the replacement of ground vehicles with aircraft. The answer is that we must integrate aircraft with the Army to the maximum degree commensurate with the capabilities of the aircraft. The object of the aircraft is not to replace, but rather to improve."[63]

Similarly, tactical air mobility doctrine also received further refinement under Taylor. The Army used Exercise SAGE BRUSH in late 1955 as an opportunity to test various tactical aspects of air mobility. This helicopter mobility test suffered some setbacks and demonstrated that further study and training were necessary.[64] The Army was nevertheless willing to incorporate more emphasis on tactical air mobility. This was evident from the 1956 Army Aviation Plan. Included under the function of "airlift for troop movement in the combat zone" was "the movement of units to execute small scale air landed operations." They were short-range actions originally of four general types. The 1958 Army Aviation Guidelines expanded this to seven general types (see Table 4-8). The 1956 Army Aviation Plan subdivided each general type further:

> (a) Actions against undefended or lightly defended objectives. There are many occasions on which the seizure of carefully selected, relatively undefended terrain will seriously compromise or even destroy the value to the enemy of his heavily defended positions. This type of action is preferable if the mission of the larger unit can be accomplished thereby.

Table 4-8
General Types of "Small Scale Air Landed Operations"

1956 Army Aviation Plan & 1957 Army Aviation Guidelines	1958 Army Aviation Guidelines
(a) Movement of troops across natural obstacles, such as rivers, swamps, lakes, canyons and escarpments. (b) Movement of troops across terrain made difficult of passage by the defensive preparations of the enemy. (c) Movement of troops to capitalize on superior mobility and speed. (d) Movement of troops to exploit shock effect of atomic fire power.	(a) Enveloping actions as a part of a larger operation. (b) Surprise assaults. (c) Seizure of critical terrain. (d) Independent actions by tactical groups designed to seek out and develop enemy forces and to direct decisive fire power against the enemy. (e) Movement of troops across natural obstacles, such as rivers, swamps, lakes, canyons, and escarpments. (f) Cover exposed flanks or wide gaps between units. (g) Movement of troops to exploit shock effect of atomic weapons.

> (b) Actions against strongly defended objectives. On the surface this may seem infeasible [*sic*] but the careful development of a technique, related to the use of atomic weapons, may prove the helicopter operation to be less expensive in human life and more remunerative tactically than conventional infantry action.[65]

These were very radical missions indeed considering that some involved operations against heavily defended positions. Moreover, these were official Department of the Army policy statements approved by Taylor.

Absorption of Air Mobility Into the Doctrinal Mainstream

The usefulness of the helicopter in assault operations had so convinced the Army high command that even the Command and General Staff College began instruction on such operations before January 1956.[66] By December 1956 the Commandant of the College created the Department of Airborne Corps and Army Aviation as part of the college's reorganization. This department was also responsible for unconventional warfare. Doctrine approved by the Department of the Army and CONARC formed the basis for its curriculum taught from the beginning of the 1958–1959 academic year. The contents of the instructions taught future division, corps, army, and unified commanders to use Army aviation in three different modes of operations. They were: in airborne operations as a supplement to Air Force troop transports, on its own in tactical and logistical support tasks, and in unconventional warfare.[67] The instruction on air mobile operations stressed that all tactical and tactical support units in the field army should be capable of participating in air mobile operations. It recognized that infantry units were most adaptable for this type of operation. However, it also included fire support and other elements. It stressed the preparation of these units to function only with air-transportable equipment.[68]

For the development of doctrine, the Department of the Army made the Infantry School the official proponent of air mobility. The Infantry branch's growing requirement for speed, mobility, and flexibility generated the largest percentage of Army aviation requirements. Colonel John Tolson said: "The Chief of Staff has said that the Department of the Army will look to The Infantry School for initiative and energetic prosecution of a training-and-development program to meet these requirements."[69] This led to an organizational change at the school to provide the staff needed. The school reorganized the Airborne Department and redesignated it as the Airborne–Army Aviation Department on 14 February 1955. (The school later renamed it the Airborne–Air Mobility Department.) Gavin contrived with Major General Robert Young (ACS, G-1, Personnel) and Major General Joseph Harper (Commandant, Infantry School) to assign Colonel Tolson as the director of this new department.[70] Tolson was an air mobility proponent.

This new department had two groups: the Airborne Training Group and the Air Mobility Group. The Air Mobility Group's mission was to develop, test, and evaluate doctrine, procedures, equipment, and organization for joint and unilateral Army operations involving air mobility. It was also to determine the requirements for the organization and equipment relating to air mobility. Finally, it was to supervise the preparation and revision of training literature.[71] The importance of the Air Mobility Group was discernible from its first product. This was the draft field manual, FM 57-35, *Army Transport Aviation Combat Operations*. In addition, it wrote the 1959 Training Circular 20-1, which CONARC and the Department of the Army approved.[72]

The Department of the Army assigned the Armor School as the major doctrinal proponent for sky cavalry. The Armor School developed sky cavalry organizations for exercises SAGE BRUSH, JUMP LIGHT, and SLEDGE HAMMER. The Armor School also partook (along with the Office, Director of Army Aviation) in developing the Aviation Center's concept of completely air-mounted sky cavalry with an air mobile ground element. The Armor branch also had an active interest in armed helicopters for cavalry missions. In 1957 there was a demonstration of one such organization at the Armor School. It was a company size unit completely air-mounted with a mixture of transport, reconnaissance, and armed helicopters, fixed wing observation airplanes, and an air mobile ground element. The Armor School was proposing to make one such company subordinate to an armored cavalry squadron (with three other ground companies).[73]

By the end of Taylor's tenure the Army high command no longer saw air mobile operations as a specialized mission. On 1 May 1959 the Department of the Army issued Training Circular 20-1 that proclaimed air mobile operations as "a normal part of land combat operations." Press reports acknowledged that this type of unilateral operation (not involving the Air Force) constituted a major if not revolutionary change. The training circular called on all combat elements to train immediately in air mobility. This training would involve using organic Army aircraft and vehicles (some of them armed) to move troops and supplies on the battlefield into battle. Training units up to company size were to use organic aircraft, while army or corps transportation units would provide lift for the training of larger units. The circular described the value of air mobile operations in economy-of-force missions, raids, anti-airborne and anti-guerrilla operations, over-obstacle assault operations, and exploitation of nuclear weapons. In addition, the circular also stated the value of air mobile operations in the seizure and retention of critical terrain, blocking or screening of enemy avenues of approach, feints, and demonstrations. It also mentioned reconnaissance and security missions, counterattack of enemy penetration, and ship-to-shore operations. The circular stated that the first objective of this training was to improve Army effectiveness by exploiting the capabilities of the air vehicles. Second, it was to gain experience that would provide data for revising current doctrine, tactics, and techniques. Third, it was to develop doctrine, tactics, and techniques to better utilize new types of air vehicles and equipment.[74]

Other doctrinal publications also reflected the increasing acceptance that tactical air mobile operations were part of the Army mainstream. In June 1958 the Army published an official version of the Infantry School's draft FM 57-35, *Army Transport Aviation—Combat Operations*. Furthermore, in April 1959 a new field manual (FM 1-100) titled *Army Aviation* appeared.[75] The organization of its chapters roughly corresponded to the six functions as approved by the 1956 Army Aviation Plan and later Guidelines. By June 1959 the Command and General Staff College devised the term "airmobile" to distinguish it from traditional parachute-type airborne operations. The term distinguished three fundamental differences between the unilateral and joint operation. First, the size of a normal airmobile operation was smaller and tactical in nature. Second, the air and ground elements were normally in the same physical area with an intimate relationship. Third, the Army commander had direct control of all major elements involved in an airmobile operation.[76] Thereafter, all Army doctrinal publications standardized their terminology to the term "airmobile." By November 1960 a new FM 57-35 appeared, now titled *Airmobile Operations*.[77] Its title gave the air mobile concept full recognition.

Internal Dissent and the Vulnerability Issue

Helicopter vulnerability was not a major issue within the Army high command and the Army aviation community. Most proponents of air mobility believed that the helicopter was not unduly vulnerable in relative terms. For example, Brigadier General Bogardus Cairns felt that helicopters were no more vulnerable than C-47 transports that carried paratroopers in World War II, or amphibious landing craft.[78] Air mobility proponents believed that tactical flying that maximized terrain features and avoided enemy concentrations would reduce exposure to enemy fire.[79] During this period the Army began to adopt the term "nap-of-the-earth" coined by Dr Doug Courtney.[80] Brigadier General Carl Hutton stressed that supportive fire on the objectives or enemy concentrations en route would also reduce helicopter exposure.[81] Major Robert Tugman and Lieutenant Colonel James Greene believed that shooting down a helicopter was not easy even when exposed to small-arms fire. They argued that "the tracers will indicate a strike on the target whereas the actual trajectory, being much like water from a garden hose, is curved and the bullets are passing below and behind the [moving] helicopter."[82] Howze and von Kann argued that armed helicopters returning suppressive fire were an additional deterrent.[83] Others such as Major Karl Patton and Lieutenant Colonel John Stockton cited the French experience in Algeria.[84] Moreover, Colonel Robert McMahon pointed out that nobody had yet organized the perfect air defense.[85]

The Army high command's view on helicopter vulnerability seems to have overruled any possibility of dissent. The 1956 Army Aviation

Plan stated that air mobile operations would subject the transport aircraft involved to "the normal hazard of combat." It stated that a well-planned action calculated to exploit enemy weakness, achieve surprise, and employ the best routes and techniques of approach would minimize these hazards. Moreover, it stated that the primary defensive measures for transport helicopter operations were low-level flight and exploitation of night and inclement weather. It saw ground fire as a primary threat only during daylight operations. It added that normal flight operations would be at altitudes below enemy radar detection.[86] The 1957 and 1958 Army Aviation Guidelines also adhered to these beliefs. In addition, the 1957 version stated that planning must include provisions for suppressive fires on enemy positions over which aircraft must fly. The 1958 version went further and said that efforts were under way to explore the feasibility of arming helicopters to provide some measure of protection from enemy small-arms fire.[87] The 1957 Guidelines also made a comment on the vulnerability of aircraft performing aerial reconnaissance missions, stating:

> The helicopter . . . is not expected to be particularly vulnerable, since it can take maximum advantage of surprise, appearing suddenly in unexpected locations; since it can hover close to cover, permitting it, if engaged by hostile fire, to drop quickly out of sight; and since the enemy gunner will find it difficult to achieve a quick hit on an object at unknown range and with no background to afford observation of strike.[88]

Despite these statements from the Army high command, helicopter vulnerability remained an emotive issue, and one that was difficult to prove or disprove. To tackle the question in a truly objective manner, the Operations Research Office (ORO) and the Combat Developments Experimentation Center (CDEC) conducted a formal aircraft vulnerability test. The Army laid out a three-year experimentation program on the vulnerability of low-flying Army aircraft to forward area ground fire. CDEC published a preliminary report entitled "Aircraft Vulnerability Experiment" on 15 June 1959. It concluded that Army aircraft were vulnerable to ground fire. Several Army agencies that reviewed the report did not agree with the methods used or the results reported.[89] As it transpired, CDEC conducted its vulnerability tests under certain controlled conditions. This restricted the speeds, altitudes, heading, and evasive action of participating aircraft, to keep the aircraft within camera range.[90] In other words, during the tests CDEC made the aircraft fly in a straight line over hostile ground fire. If the ground troops acquired the target within their sights (captured by camera), then it would determine the aircraft as shot down. CDEC did not allow evasive flying, nor did it allow preflight planning to avoid known hostile areas. The results of the test therefore had little validity.

Moreover, the helicopter vulnerability issue has intrinsic links with the issue of conservatism. When dealing with this issue, one must distinguish between skepticism and vested bureaucratic interests. Sceptics are not able to visualize the possibilities of a new way of warfare or are uncertain about the degree of risk involved. For air mobility, helicopter vulnerability would be a central issue of uncertainty. Bureaucracies with vested interests in the status quo might use helicopter vulnerability to resist change and thus preserve and protect their preeminent position and budgets.

Documented instances of outright opposition or conservative behavior were rare. Instead, suggestions that conservatism existed tended to come from air mobility proponents. Some exercise or test reports did raise the issue of vulnerability and had some dissenting views towards Army aviation policies.[91] However, there was evidently little serious opposition to a radical air mobility doctrine within the top Army echelons. The internal opposition that existed was insufficient to affect seriously the Army high command's funding allocation and program support given to Army aviation.

There were several other possible factors why there was a lack of serious internal Army dissent or discord over the general policy of investing in air mobility. While the creation of the Association of the United States Army (AUSA) was for external political purposes, it was also partly to deal with the Army's lack of "inner unification." The Army high command used the AUSA annual symposium to disseminate their radical views. The intense interservice conflicts also probably reduced the intraservice conflicts. The Army perhaps saw air mobility as something desirable to possess vis-a-vis the Air Force even if it had a limited usefulness. The Army was also a more eclectic group with many long and differing historical traditions. It therefore does have greater tolerance for diverse groups such as a growing Army aviation community. Moreover, air mobile operations were not radically dissimilar from airborne operations. The only difference is in the operation of flying machines.[92]

THE INEFFECTIVENESS OF EXTERNAL RESTRICTIONS

During this period the Army described the Air Force position towards Army aviation developments as unmistakably "one of determined and sometimes violent opposition."[93] This was a part of the intense interservice rivalry. Instead of the two services delineating aviation functions between themselves, the Department of Defense began to impose its will on this matter. Secretary of Defense Charles Wilson issued a memorandum on 26 November 1956, entitled "Clarification of Roles and Missions to Improve the Effectiveness of Operation of the Department of Defense." Besides aviation issues, it also dealt with other interservice controversies. Reissued as Department of Defense Directive No. 5160.22 on 18 March 1957, it replaced the 1952 Pace-Finletter agreement (see Table 4-9). The title of this directive was "Clarification of Roles and Missions of the Departments of the Army and the Air Force Regarding Use of Aircraft."[94] This directive, dealing only with the aviation issue, defined the Army aviation functions in a similar but more detailed manner. It included restrictions on the weight of Army aircraft and their operational area. Otherwise the two documents were essentially identical. Senior Army officers, press reports, and Army lobby groups denounced both. For example, General Maxwell Taylor wrote in his 1959 book:

> As technical improvements in weapons and equipment offered the Army the possibility of escaping from dependence upon the Air Force, the latter has vigorously resisted these efforts and has succeeded in obtaining the support of the Secretary of Defense in imposing limitations on the size and weight of aircraft procured by the Army, on the ranges of Army missiles, and on the radius of Army activities in advance of the front line of combat.[95]

Similarly, Gavin also made public his discontent with the administration restrictions after he retired in 1958:

> At the time the memorandum was issued, it was made clear to the Army that it was fortunate to get off so lightly, that serious consideration was being given to more restrictive limitations. Hence at the very time when it should have been encouraged to go ahead in these new fields, it was forbidden to do so. The effect

> of the memorandum was to close the vast area between the Army's light air vehicles and the large vehicles of the Air Force to further exploration and research.[96]

General Gavin, testifying to Congress earlier, detailed the exact effects of the 1956 restrictions:

> The weight limitations on Army aircraft, if strictly applied, will unquestionably prevent the Army from developing the aircraft necessary for fulfillment of its mission. To quote one example, the Army must have integral aircraft capable of moving reasonable numbers of men or quantities of materiel within the combat zone. The requirements for this sort of transport are such that they cannot be economically fulfilled with fixed wing aircraft weighing less than 5,000 pounds empty. Strict interpretation of the directive will therefore force the Army to perform these tasks with helicopters. In any given payload class a helicopter generally costs at least twice as much as an airplane initially, and five times as much to maintain. The Army has further indications that the heavier categories of conventional helicopters, that is above 3 tons payload, will be extremely difficult to maintain at any cost under field conditions. The Army has therefore felt it essential to supplement helicopters to the maximum extent possible with short takeoff and landing and vertical takeoff and landing airplanes and has pioneered research in airplanes of this type to make this possible.[97]

During a 1960 testimony to the House Military Appropriations Committee, Lemnitzer also hit out against the aircraft restrictions imposed by the Department of Defense. He said that these limitations had inhibited efforts in research and development. He added: "Any arbitrary definition of the 'combat zone' which restricts the operation and employment of [Army] equipment is objectionable in principle."[98] Lemnitzer was struggling hard to get the inhibiting restrictions withdrawn.[99] The Army Director of Research and Development, Richard Morse, also stated in late 1960 that the Army was pressing the Department of Defense to repeal the aircraft limitations.[100]

Table 4-9
Department of Defense Directive No. 5160.22, 18 March 1957

Definition:
. . . the combat zone is defined as extending not more than 100 miles forward of the general line of contact between US and enemy ground forces. Its extension to the rear of the general line of contact will be designated by the appropriate field commander, and normally extends back of the front lines about 100 miles.
Army aviation functions:
a. *Command, liaison, courier and communications* – This includes aerial wire-laying and aviation to assist in the direction, coordination and control of Army forces in the field. b. *Observation, visual and photographic reconnaissance, fire adjustment, and topographic survey* – This includes aerial observation to amplify and supplement other Army methods of observation for the purpose of locating, verifying and evaluating targets, adjusting fire, terrain study, or obtaining information on enemy forces, complementing that obtained by air reconnaissance agencies of the other Services; this includes limited aerial photography incident to these purposes. c. *Airlift of Army personnel and materiel* – Transportation of Army supplies, equipment, personnel, and small units within the Army combat zone in the course of combat and logistical operation. Includes the movement of small units to execute small-scale air-landed operations, the movement of reserves, and the shifting or relocation of small units and individuals within the combat zone as the situation may dictate. Includes expeditious movement of critially needed supplies or equipment, or both, within the combat zone, supplementing the ground transportation system operating within the field army. Does not include the execution of joint airborne operations. d. *Aeromedical evacuation* – Aeromedical evacuation within the Army combat zone to include battlefield pickup of casualties (except those from an airhead or airborne objective area which is supported by Air Force airlanded logistical support), air transport to initial point of treatment and any subsequent moves to hospital facilities within the Army combat zone.
Air Force logistical air support functions:
a. Airlift of Army supplies, equipment, personnel and units from exterior points to points within the Army combat zone. b. Airlift for the evacuation of personnel and materiel from the Army combat zone. c. Airlift for the air movement of troops, supplies, and equipment in the initial and subsequent phases of airborne operations. d. Aeromedical evacuation from Air Force operating locations within the combat zone through Air Force casualty staging units to hospital facilities outside the combat zone; and the aeromedical evacuation from an airhead or an airborne objective area where airborne operation includes air landed logistic support by Air Force.

However, neither the restrictions nor the Air Force opposition proved to be an obstacle for the Army achieving its aviation goals. This was apparent from the way the Department of Defense implemented or interpreted the restrictions and how it arbitrated over other disputed issues between the two services. The Department of Defense had only to keep the defense budget within government means. This was the ultimate determinant of the restrictions imposed on the Army.[101] Indeed, Wilson's successors Neil McElroy (9 October

Table 4-9 continued
Department of Defense Directive No. 5160.22, 18 March 1957

Limitations for Army aviation:
1. Fixed wing aircraft, convertiplanes, and vertical/short take-off and landing aircraft will have an empty weight not to exceed 5,000 pounds. Rotary wing aircraft will have an empty weight not to exceed 20,000 pounds. Specific exceptions to weight limitations for specific aircraft for specific purposes may be granted by the Secretary of Defense after consideration of US Army requirements and appropriate US Air Force functions and capabilities.
2. The provision of a limited airlift capability within the US Army Aviation Program shall not serve as a basis for increasing or decreasing US Air Force forces necessary to support or protect the US Army airlift forces. Provision of this limited airlift capability will apply only to small combat units and limited quantities of materiel to improve local mobility, and not the provision of an airlift capability sufficient for the large-scale movement of sizeable US Army combat units which would duplicate the mission of the US Air Force.
3. As limited Army aviation airlift capability becomes available to active Army forces, provision should be made for compensating reductions in other forms of Army transportation designed to operate within the combat zone.
4. The US Army Aviation Program will not provide for aircraft to perform the following functions:
a. Strategic and tactical airlift, as outlined [in the Air Force logistical air support functions] above.
b. Tactical reconnaissance.
c. Interdiction of the battlefield.
d. Close combat air support.
5. The US Army will not maintain unilateral aviation research facilities, but will confine itself to development and determination of specific requirements peculiar to Army needs, to evaluation of proposals, and to user testing of equipment. The US Army will make maximum use of US Air Force and US Navy aircraft research and development facilities. The US Air Force and the US Navy will be responsive to US Army needs in such research activities on a reimbursable basis.
6. The US Army will use existing types of US Navy, US Air Force or civilian aircraft when they are suitable, or may be suitably modified, to meet Army requirements, rather than attempt to develop and procure new types.

1957 to 1 December 1959) and Thomas Gates (2 December 1959 to 20 January 1961) did give some thought to removing the restrictions on Army aircraft entirely.[102] It led the Army to view the Department of Defense as "reservedly favorable" towards Army aviation and air mobility.[103]

The contents of the 1956 memorandum and the 1957 directive were not as restrictive as some of their critics made out. For example, the Air Force tried to stop the Army from using its transport helicopters behind enemy lines for assault operations. It saw the Army use of helicopters to land small units in enemy territory as an infringement of its responsibility for providing airlift for airborne operations. Superficially, the restrictions did state that the Army must not duplicate this Air Force function. However, it clearly allowed the

Army the right to execute air mobile operations over enemy-held territory. The 1952 Pace-Finletter agreement defined the combat zone as having a normal "depth" of 100 miles, that is, stretching rearward from the front lines. The new rules redefined the combat zone as extending 100 miles forward and rearward of the general line of contact. Moreover, it stated that the Army airlift of personnel and materiel included "the movement of small units to execute small-scale air-landed operations" within this combat zone. The only restriction was on the size of such operations, which must not be "large-scale" or involve "sizeable Army combat units."

Similarly, the Air Force contested the Army's right to execute sky cavalry operations, which operated behind enemy lines to gather information. While the 1956 memorandum and the 1957 directive did not mention sky cavalry, the Department of Defense ruled afterward in the Army's favor. For example, Deputy Secretary of Defense Reuben Robertson said publicly that the memorandum would not bar the Army from developing the sky cavalry–type, aerial reconnaissance units. The press, reporting this on 1 December 1956, saw this as a clear victory for the Army.[104] An Army officer concluded in an April 1957 article that Wilson granted authority to develop and employ sky cavalry units since the memorandum did not mention it.[105] Since Army experiments and development of helicopter armaments (an important facet of sky cavalry) continued without interference, this was quite likely true. Indeed, the Department of Defense approved the Army's desire to adopt armed helicopters.[106] Moreover, it earlier did not interfere with the Army's experiments with armed fixed wing airplanes. However, there were no plans for implementation perhaps because of its controversial nature.

The memorandum and directive also did not deal directly with the issue of a higher performance observation airplane. The Department of Defense, however, accepted in principle the Army's insistence for such an airplane. What it did was to impose procurement rules that resulted in an aircraft type different from what the Army originally envisaged (using turbo-prop rather than jet engines).

The weight limit seems to be a particularly contentious issue. Prior to the 1956 memorandum the Army Secretary and the Chief of Staff were applying to the Department of Defense for a cargo airplane, against the opposition of the Air Force. The new rules not only retained the 5,000-pound weight limit on fixed wing aircraft, it also imposed a limitation of 20,000 pounds on helicopters. Furthermore, the 5,000-pound limit also applied to convertiplanes and vertical/short take-off and landing aircraft exempt from the 1952 agreement. However, there was clearly a concession on the weight issue by having an allowance for exceptions, with a formal procedure established to do so. Even the Deputy Chief of Staff for Military Operations stated that the procedure for obtaining such exceptions was satisfactory. General Eddleman's memorandum to the Army Chief of Staff concluded that Wilson's memorandum did not impair the Army's ability to carry out its assigned roles and missions.[107] Moreover, General Howze remarked that "the Secretary of Defense has indicated clearly that his intent is not to limit the size of Army aircraft categorically, but rather to require thorough justification for large aircraft."[108]

Indeed, when Wilson issued the 1956 memorandum, he had already approved one exception. The memorandum stated that "the Secretary of Defense has just approved the purchase by the Army of five DeHavilland DHC-4 airplanes, 'Twin-Otter,' for test and evaluation and is giving consideration to another project involving a plane in the development stage." The DHC-4 Twin-Otter was then in the design stage and destined to enter service with the Canadian forces as well. It eventually turned into the AC-1 (CV-2) Caribou medium tactical transport aircraft with a weight of 16,920 pounds empty. The one under consideration (which Wilson eventually approved) was the AO-1 (OV-1) Mohawk, weighing 9,937 pounds empty. This was the turbo-prop higher performance observation airplane then under joint development with the Marine Corps.

The Department of Defense also understood the Army's need to consolidate some dual-service administrative responsibilities from the Air Force. The size and nature of Army aviation had changed beyond recognition since the interservice agreements made in 1945 and 1950. (See Chapter 1 for the 3 August 1945 and 23 March 1950 agreements.) The Department of Defense therefore did not challenge the Army's assumption of unilateral responsibility over depot supply and maintenance, and primary flight and mechanic training.[109] The 1956 memorandum and the 1957 directive stated that the Army may employ its aircraft in training. Separately, the Army, the Air Force, and the Department of Defense all agreed and coordinated the changes to depot supply and maintenance. The Department of Defense did, however, continue to prohibit the Army from having its own aviation research and development facilities. The Army was to continue making maximum use of Air Force or Navy facilities (a new addition). The Army nevertheless increasingly assumed more responsibilities over research, development, and procurement, making the process more streamlined. In short, the Department of Defense's tacit approval or noninterference was instrumental to the considerable advance in Army aviation during this period.

NOTES

1 The United States Government's fiscal practice for this period under discussion requires an explanation. Congress assigns money to each service annually. This is the "new appropriations." However, for the period under discussion, the service does not need to spend this amount within the fiscal year. It can "incur obligations," that is, designate money for contracts. The Army can sign a contract but it does not need to pay the entire sum at once. It is more usual to have a down-payment, progress payment(s), and final payment at the time of final delivery. This final payment may occur over 12 months later. The actual sum paid out for a fiscal year is the "expenditure." The "carryover" money is the difference between new appropriations and obligations (the unobligated carryover), and the obligations less the amount spent. The budget for Army aviation given throughout the book concerns the obligated amount, and is therefore only one aspect of the fiscal picture.

2 House of Representatives, *Army Appropriations FY 59 Part 1*, pp. 394–95.

3 (1) General Clyde D. Eddleman, Address to AUSA Army Aviation Symposium, Fort Rucker, Alabama, "Planning for Army Aviation," 23 March 1961. (2) "Policy-Bound Army Hunts Air Tools for Atomic Warfare," *AW*, 25 February 1957, p. 88.

4 Major General Frank S. Besson, Jr, Chief of Transportation, Department of the Army, "TC Seeks to Improve Mobility Capabilities," *AT*, 30 April 1960, Special Section, p. 40.

5 (1) Bacevich, p. 103. (2) "Army Molds Air Weapon Arsenal," *AW*, 31 October 1955, p. 17. (3) Claude Witze, "Army Budget Stresses Planes, Missiles," *AW*, 16 January 1956, p. 31. (4) Richard Sweeney, "Gavin: Army Needs Planes, Missiles," *AW*, 8 October 1956, p. 29.

6 DoD, *Semiannual Reports, Jan.–June, 1954*, p. 122.

7 Weinert, p. 16.

8 White, p. 23.

9 (1) "Secretary Stevens Says Aircraft Lack Bars Army from Crossing Mobility 'Threshold,'" *ANAFJ*, 25 September 1954, p. 97. (2) "'Copter Called 'Greatest Single Advance,'" *ANAFJ*, 25 September 1954, p. 97.

10 (1) "Rapid Mobility, Air Transportation, Basis of Army Planning, Says Asst. Sec.," *ANAFJ*, 9 October 1954, p. 151. (2) "Symposium Attacks Supply, Maintenance Problems of Expanding Army Aviation," *ANAFJ*, 3 November 1956, p. 14.

11 DoD, *Semiannual Reports, Jan.–June, 1955*, pp. 88, 123–24.

12 (1) Claude Witze, "Army Creates Aviation Division," *AW*, 21 March 1955, p. 20. (2) "Service Careers of Generals Howze and Hutton," *ANAFR*, 26 November 1955, pp. 17, 24.

13 (1) Weinert, p. 42. (2) "Army's Aviation Division," *AA*, 5 December 1955, p. 18. (3) "The Army's Month: D/A Staff Reorganization," *The Army Combat Forces Journal*, January 1956, pp. 12–13.

14 Eddleman, "Planning for Army Aviation."

[15] (1) Colonel R. R. Williams, "Ground Forces 'Obsolete Term' Says Army Aviation Pioneer at Rucker Center," *ANAFJ*, 28 July 1956, p. 11. (2) "Army Aviation Board Covers All Phases of Industry and Other Services to Promote Utilization of Air by Soldiers," *ANAFJ*, 9 March 1957, p. 11.

[16] (1) Brigadier General Carl I. Hutton, Commanding General, The Army Aviation Center, "Organic Aviation Can Boost Army Power a Thousand Fold, Writes Center's CG," *ANAFJ*, 8 October 1955, pp. 1, 28. (2) Bob Williams, "Salute to Fort Rucker," *ANAFR*, 26 November 1955, pp. 17–24. (3) "Ft. Rucker Shows Army Way to Use Aviation for Modernized Forward Push," *ANAFJ*, 2 March 1957, pp. 1, 4.

[17] Weinert, pp. 9–10.

[18] Colonel John J. Tolson, "Wings for Infantry," *Infantry School Quarterly*, July 1956, pp. 12–21.

[19] Headquarters, The Army Aviation Center, Camp Rucker, Alabama, Memorandum for distribution, "Letter from General Howze, Army Aviation Division, G3, Department of the Army," 14 May 1955.

[20] DA, "1956 AAP," Annex 2, Appendix A, Inclosure 1.

[21] "Army Has 6-Year Aviation Program, 28 Divisions Ideal, Says Gen. Taylor," *ANAFJ*, 18 February 1956, p. 3.

[22] DA, "1956 AAP," p. 1; Annex B, Appendices A to F.

[23] Taylor, "The Army Rewrites the Book."

[24] House of Representatives, *Army Appropriations FY 59, Part 1: Overall Policy Statements*, p. 394.

[25] "Army Planning 50% Increase in Pilots and Aircraft; Replace All Planes by 1960," *ANAFJ*, 7 September 1957, p. 2.

[26] General Maxwell D. Taylor, Army Chief of Staff, "The Army Needs Mobility," *NDTJ*, November–December 1955, p. 52.

[27] (1) House of Representatives, *Army Appropriations FY 58 Part 1*, pp. 100–101. (2) "Army Wants Own Integral Aviation," *ANAFJ*, 19 November 1955, p. 2. (3) Wilber M. Brucker, Secretary of the Army, "This is a Modern Army," *Army*, September 1959, pp. 27–28. (4) Wilber M. Brucker, no title, *USAAD*, June 1957, p. 2.

[28] "Army Aviation Faces Expansion; Pilot-Plane Boost Nears," *ANAFJ*, 24 September 1955, p. 20.

[29] DoD, *Semiannual Reports, Jan.–June, 1956*, p. 83.

[30] Bacevich, pp. 22–24, 75, 145.

[31] Taylor, *Swords and Plowshares*, p. 171.

[32] JATB, "1953 Conference," Appendix A2, p. 57.

[33] (1) "Army Will Call on Industry Help for Limited War Aviation Design Concepts," *ANAFJ*, 14 November 1959, p. 2. (2) "Army Plans 1970 Aviation Program," *ANAFJ*, 28 November 1959, p. 26.

[34] (1) Aviation School, "1960 Instructors' Conference," pp. 17, 19, 20. (2) "Double Plane Money, Says L/G Trudeau," *ANAFJ*, 22 October 1960, p. 28.

[35] "Army Sparks Aviation Build-Up to Ease Airlift Gap; Hits DoD Restrictions," *ANAFJ*, 20 February 1960, p. 23.

[36] (1) Lieutenant General Arthur G. Trudeau, Chief of Research and Development, Department of the Army, "Toward a More Mobile Army,"

NDTJ, November–December 1958, pp. 46–47, 84–85. (2) Trudeau, "Toward a Mobile Army," pp. 37, 60–63. (3) "Army Accents Air to Aid Mobility," *ANAFJ*, 6 August 1960, pp. 1, 3. (4) "Double Plane Money, Says L/G Trudeau," p. 28.

37 "Army Aviation Seeks Faster, More Powerful Planes," p. 11.

38 "Army Air Arm Gets Big Building Boost; Rucker Work Alone Totals $7.3 Million," *ANAFJ*, 14 July 1956, p. 11.

39 "Army Sparks Aviation Build-Up to Ease Airlift Gap," p. 23.

40 "Why Doesn't the Army 'Make a Fight' to Kill Restrictions on Air Arm?" *ANAFJ*, 23 April 1960, pp. 1, 20.

41 "Army's Air Arm Needs Budget Help; Committee Reports Shortages," *ANAFJ*, 2 July 1960, p. 4.

42 (1) Bergerson, p. 192, note 1. (2) "Army Aviators to Meet," *ANAFJ*, 9 May 1959, p. 7. (3) "Army Aviation President Wastes No Time; Concern Expressed Over Wing 'Clipping,'" *ANAFJ*, 13 June 1959, pp. 2, 4.

43 Major Robert Winkler, "Height of Battle," *Infantry School Quarterly*, January 1955, p. 99.

44 "Army Orders New S-58 Copters," *AW*, 13 July 1953, p. 26.

45 The Transportation School, Special Text ST-55-21-1, "Transportation Helicopter Units," Fort Eustis, Virginia, June 1952, p. 8, paragraph 9.

46 United States Army Infantry School, Memorandum to Staff and Faculty, Infantry School, "Transport Helicopter Doctrine," n.d. [c.1954].

47 Weinert, p. 119.

48 (1) Weinert, pp. 58–59. (2) "Army Activates First Fixed Wing Plane Unit," *AW*, 13 June 1955, p. 18. (3) "Army to Add Air Support Companies," *AFT*, 4 June 1955, p. 29.

49 (1) United States Army Infantry School, "Draft Manuscript, Department of the Army Field Manual, FM 57-35, Army Transport Aviation Combat Operations," Fort Benning, Georgia, June 1955. (2) Major James S. Douglas, "Exercise Lift," *The Army Combat Forces Journal*, November 1954, pp. 20–23.

50 Brigadier General Carl I. Hutton, "Without Trumpet and Drums: Recollections of 36 Months with the Army Aviation School, June 1954–June 1957," USAAVNC, Technical Library, Fort Rucker, Alabama, n.d. [c.1957], p. 148.

51 Captain John C. Buchanan, "The Vertical Patrol," *Armor*, September–October 1955, pp. 4–7.

52 DoD, *Semiannual Reports, Jan.–June, 1955*, p. 88.

53 (1) Yount, "Air Transportation," p. 64. (2) Brigadier General Richard D. Meyer, "The Next Decade in Army Aviation," *USAAD*, August 1958, p. 30. (2) "Army Has Four Priority Aircraft, General Meyer Tells Aviation Writers," *ANAFJ*, 23 May 1959, p. 7.

54 General Maxwell D. Taylor, Chief of Staff, US Army, "The Army's Need for Mobility," *NDTJ*, November–December 1957, p. 77.

55 Major General Hamilton H. Howze, "Combat Tactics for Tomorrow's Army," *Army*, October 1957, p. 29.

56 (1) Weinert, pp. 62, 64–67. (2) Infantry School, "1956 Instructors' Conference," p. 132.

57 Infantry School, "1957 Instructors' Conference," pp. 40, 46.

58 United States Army, Headquarters Exercise STRONG ARM, "Logistical Support of Armor by Air Force and Army Aircraft, Report of Troop Test," Fort Hood, Texas, 1–9 May 1958.

59 (1) United States Army Combat Development Experimentation Center, "Pentana-Type Companies in Mobile Operations (U), Interim Evaluation Report, Volume I: Basic Report of Experiment, Volume II: Annexes, Volume III: Detailed Plan of Experiment," Fort Ord, California, 1 March 1958, p. 8, and Abstract. (2) United States Army Combat Development Experimentation Center, Training Text 58T1-1, "Integrated Combat Group (PENTANA-Type) (U)," Fort Ord, California, March 1959, pp. 9, 87, 88.

60 Downing, pp. 48–53.

61 (1) Major General Paul F. Yount, Chief of Transportation, "Transportation for Tomorrow's Army," *AID*, February 1958, pp. 12–15. (2) "Army Aviation Today," *AA*, 10 March 1958, p. 32.

62 Claude Witze, "Military Holds Key to Copter Future," *AW*, 14 March 1955, p. 296.

63 Major General Paul F. Yount, Army Chief of Transportation, "Transportation for Peace," *NDTJ*, July–August 1957, p. 61.

64 Weinert, pp. 123, 125.

65 DA, "1956 AAP," Annex 1, Appendix B, p. 1–2.

66 Lieutenant Colonel Stuart M. Seaton, Faculty, Command and General Staff College, "The Helicopter in Early Link-Up Operations," *MR*, January 1956, pp. 33–42.

67 Colonel Robert E. McMahon, Faculty, US Army Command and General Staff College, "Keeping Pace with the Future—Air Mobility for Army Forces," *MR*, July 1958, pp. 49–62.

68 Colonel Robert E. McMahon, Faculty, US Command and General Staff College, "Airmobile Operations," *MR*, June 1959, p. 34.

69 Tolson, "Wings for Infantry," p. 17.

70 Tolson, *Airmobility*, p. 5.

71 (1) Tolson, "Wings for Infantry," p. 18. (2) For the role of the Infantry School on doctrinal formulation, see Colonel John J. Tolson, Director, Airborne–Army Aviation Department, The Infantry School, "Air Mobility of Army Striking Force Studied by Infantry School Aviation Dept.," *ANAFJ*, 1 October 1955, p. 11.

72 Monte Bourjaily, Jr, "Training to Stress Airmobility," *AT*, 23 May 1959, p. 1.

73 Major Robert F. Tugman and Lieutenant Colonel James S. Greene, Jr, "An Aero-Cavalry Concept," *Armor*, May–June, 1957, pp. 44–47.

74 (1) Bourjaily, pp. 1, 28. (2) "Airmobility Training Broadened in Army," *AFT*, 8 May 1959, p. 18.

75 (1) Department of the Army, Field Manual, FM 57-35, *Army Transport Aviation—Combat Operations* (Washington 25, DC, 10 June 1958). (2) Department of the Army, Field Manual, FM 1-100, *Army Aviation* (Washington 25, DC, 3 April 1959).

76 McMahon, "Airmobile Operations," p. 32.

77 Department of the Army, Field Manual, FM 57-35, *Airmobile Operations* (Washington 25, DC, 2 November 1960).

78 "Army Aviation Leader Tells Prospect for Added Punch to Pentomic Divisions," *ANAFJ*, 20 September 1958, p. 3.

79 (1) Infantry School, "1957 Instructors' Conference," pp. 27–28. (2) Oswalt, "Why and How Army Aviation Arms for Battle," p. 42.

80 Colonel John D. Edmunds, Assistant Commandant, The Army Aviation School, "US Army Participation in Aircraft Instrumentation Development," *Journal of the American Helicopter Society*, April 1957, p. 30.

81 Brigadier General Carl I. Hutton, Commanding General, 8th Infantry Division Artillery, "Helicopter-Borne Operations," *USAAD*, October 1958, pp. 5–8.

82 Tugman and Greene, p. 45.

83 (1) Major General Hamilton H. Howze, Director of Army Aviation, ODCSOPS, "Notes from the Pentagon: Meditations on Helicopter Vulnerability," *USAAD*, October 1957, p. 3. Hereafter cited as Howze, "Meditations on Helicopter Vulnerability." (2) von Kann, "Mobility Now," pp. 40–42.

84 (1) Major Karl S. Patton, Chief of Doctrine Division, Combat Development Office, US Army Aviation School, "Aerial Combat Reconnaissance Company Doctrine and Tactics," *USAAD*, August 1959, p. 13. (2) Lieutenant Colonel John B. Stockton, "Cavalry, 1960," *Armor*, January–February 1960, p. 9.

85 McMahon, "Airmobile Operations," p. 33.

86 DA, "1956 AAP," Annex 1, Appendix B, p. 2.

87 (1) Department of the Army, Office of the Deputy Chief of Staff for Military Operations, "Army Aviation Guidelines for the Development of Doctrine and Organization Through FY 1961," Washington, DC, 14 February 1957, p. 4; Annex 1, Appendix B, p. 2 Hereafter cited as DA, "1957 AAG." (2) DA, "1958 AAG," Annex 1, Appendix B, pp. 2–3.

88 DA, "1957 AAG," Annex 1, Appendix D, p. 2.

89 (1) United States Army Aviation School, Combat Developments Office, Office, Ordnance Corps Liaison Officer, Fort Rucker, Alabama, Memorandum to Commanding General, Ordnance Weapons Command, Rock Island, Illinois, "Ordnance Corps Liaison Officer's Report," 7 July 1959. (2) United States Army Infantry School, Headquarters, "Developmental Objectives for Army Aviation 1965–1970," Fort Benning, Georgia, 4 December 1959, Annex I, pp. 13, 16. Hereafter cited as Infantry School, "DOAA 65–70."

90 Infantry School, "DOAA 65–70," Annex I, p. 24.

91 Weinert, pp. 97, 122–23, 125–26.

92 (1) Halperin, p. 35. (2) Huntington, pp. 395, 407.

93 Goodhand, "Army Aviation Policy," p. 1.

94 Goldberg, *Defense Documents*, pp. 312, 315.

95 Taylor, *The Uncertain Trumpet*, pp. 168–69.

96 Gavin, *War and Peace in the Space Age*, p. 157.

97 Senate, *DoD Appropriations FY 58*, p. 520.

98 "Army Sparks Aviation Build-Up to Ease Airlift Gap," p. 23.

99 "Why Doesn't the Army 'Make a Fight'?" p. 1.

100 "Army Research Boss Asks DoD to Lift Restrictions on Weapons; Pilot Cut Hit; Warns Against 'Stagnation' of 10-Year Program," *ANAFJ*, 12 November 1960, p. 3.

101 "Fulfilling Roles and Missions," *ANAFJ*, 27 March 1957, p. 10.

102 "Why Doesn't the Army 'Make a Fight'?" pp. 1, 20.

103 Goodhand, "Army Aviation Policy," p. 1.

104 "AF-Navy Get Missile Green Light; Shaken Army Looks to Bright Spots," *ANAFJ*, 1 December 1956, p. 3.

105 Colonel Frank G. Forrest, Aviation Officer, Sixth US Army, "Effect of Clarification of Roles and Missions on Army Aviation," *USAAD*, April 1957, p. 25.

106 Bourjaily, p. 1.

107 (1) Lieutenant General C. D. Eddleman, Deputy Chief of Staff for Military Operations, Memorandum to the Chief of Staff, US Army, "Clarification of Roles and Missions, SecDef Memorandum of 26 November 1956," n.d. (2) Deputy Chief of Staff, Summary Sheet to Chief of Staff, "Analysis of SecDef's Memo, dtd 26 Nov 56, Subject: 'Clarification of Roles and Missions to Improve the Effectiveness of Operation of the DOD,'" 7 December 1956.

108 Howze, "Last Three Years," p. 46.

109 (1) Weinert, pp. 144–46, 184–95, 219–20. (2) J. K. Gerdel, "Army Aviation Depot Plan," *NDTJ*, September–October 1958, pp. 60, 85–87.

SOURCES FOR TABLES

4-1. (1) DoD, *Annual Reports FY 59* and *FY 60*, pp. 30; 34. (3) Bureau of Budget, *FY 50* to *FY 52*, pp. A5, A5, A8 respectively.

4-2. House of Representatives, *Army Appropriations FY 56* to *FY 61*, pp. 1451, 1460; 1450; 1353, 1356; 386–87; 149; 3, 64 respectively. The figures represent direct obligations with no reimbursement amounts. For FY 54 this may be significant, as the reimbursable obligations for major equipment stood at $595,889,970. The total reimbursable obligations amounted to $919,150,769 and pushed the total obligations to $2,039,930,615. In the first six months of FY 54 alone the Army obligated $97.8 million for aircraft. See "Army Aircraft Buying to Double in 1955," *AW*, 26 April 1954, p. 16.

4-3. Bureau of Budget, *FY 53* to *FY 67*, pp. 596, 619–20; 556, 573–74; 482, 506–7; 506, 530–32; 507, 532–33; 487, 509–11; 435, 438, 456–57; 451, 455, 487–88; 441, 444, 471–73; 478, 480, 516–18; 167, (Appendix) 252–54, 333; 190, (Appendix) 273–75, 355; 205, (Appendix) 266–69, 343; 218, (Appendix) 295–97; 232, (Appendix) 309–11 respectively. The item "Procurement and Production, Army" was only created from FY 51 onward. Before this the procurement and production budget was submerged under various other accounts. The FY 52 brought forward and appropriation amounts were therefore transfers from other accounts. Note that carryover amounts for FY 54 to FY 55 differed in Bureau of Budget, *FY 56* and *FY 57*, FY 57 figures used.

4-4. (1) Bureau of Budget, *FY 59* to *FY 62*, pp. 433; 448; 438; 486. (2) DoD, *Semiannual Reports, Jan.–June, 1957*, *Jan.–June, 1958*, and *Annual Reports FY 60*, pp. 98; 113; 6. (3) Olson, p. 165.

4-5. Bureau of Budget, *FY 53* to *FY 61*, pp. 624–25; 577–78; 502–4; 525; 505; 505; 455; 499; 484 respectively.

4-6. (1) House of Representatives, *Army Appropriations FY 57*, p. 336. (2) DA, "1956 AAP," Annex 1, Appendix B, pp. 2–3; Annex 1, Inclosure 2, Chart 2. (3) DA, "1957 AAG," p. 4; Annex 1, Inclosure 1, Table V and VI. (4) DA, "1958 AAG," Annex 1, Appendix G, Table 3 and 4.

4-7. (1) DA, "1956 AAP," Annex 2, Appendix A, Inclosure 1. (2) DA, "1957 AAG," Annex 2, Appendix A, Inclosure 2. Cargo helicopter assets for 1956 includes two medium cargo helicopters. (3) DA, "1958 AAG," Annex 2, Appendix A, Inclosure 2. The 1958 figures for observation airplanes and cargo helicopters included 4 medium observation airplanes and 48 medium cargo helicopters. (4) Aviation School, "1960 Instructors' Conference," p. 15. The 1959 figures for utility and cargo helicopters included 9 HU-1 helicopters and 72 H-37 respectively.

4-8. (1) DA, "1956 AAP," Annex 1, Appendix B, pp. 1–2. (2) DA, "1957 AAG," Annex 1, Appendix B, pp. 1–2. (3) DA, "1958 AAG," Annex 1, Appendix B, pp. 1–2.

4-9. Goldberg, *Defense Documents*, pp. 312–15.

L-4 Piper Cub (Grasshopper)

L-5

L-17 Navion

L-20 Beaver

L-23

U-1 Otter

OV-1 Mohawk

CV-2 Caribou

H-13

H-19

H-21

H-34

H-37

YH-16

XH-17

XV-3 tilt-rotor convertiplane

5

PUTTING THEORY INTO PRACTICE

Helicopter technology was a major limitation on air mobility implementation plans during the Eisenhower era, more so than any other factor. The transport helicopter deficiencies led to a shift in emphasis towards fixed wing transports and more light transport helicopters. Apart from transforming the aviation force structure plans, it also determined the direction of Army aircraft research and development. While the doctrinal development of air mobility progressed well during the fifties, the Army had a less than adequate ability to execute any large-scale or widespread air mobile operations. Problems with indoctrination and training meant that the Army was ill-prepared operationally. One possible solution was the creation of dedicated air mobile combat units. However, the Army high command was not yet ready to introduce such units. The development of sky cavalry concepts into actual capability also ran into difficulties. This delayed the introduction of sky cavalry units until the next decade. In short, while the theory of air mobility was advancing, the Army was finding it difficult to carry it out in practice.

OVERCOMING TRANSPORT HELICOPTER DEFICIENCIES

There were three main problems associated with transport helicopters during the Eisenhower period: inadequate production quantities, high initial procurement costs, and high running costs. High running costs broke down into two further problems: excessive maintenance requirements, and problems with spare part supply and costs. Army helicopter operations in Korea had already uncovered the problem of an inadequate maintenance and supply system to support distant field use.[1] Low availability was also a problem. In the early years the Army had an ultimate objective of achieving 75 percent availability, although it was expecting 67 percent availability. However, actual operations in Korea proved that the availability rate was only 57 percent.[2] This problem was entirely the result of excessive maintenance requirements and problems with spare part supply. Solving the maintenance and spare part supply problems would indirectly deal with low availability. The bases for these deficiencies were all the same: the inherent technical complexity and infancy of helicopter technology. Both General Howze and Colonel William Bunker[3] warned that the excessive running costs and low availability would adversely affect the Army's procurement plans.[4]

Inadequate Production Quantities

The problem with inadequate production quantities originated from the recommendations of the 1952 Materiel Requirements Review Panel. The main focus was on a family of light ($1\frac{1}{2}$–ton), medium (3-ton), and heavy (5-ton) cargo helicopters. Utility helicopters were a minor requirement. Technical complexity, however, increased as helicopters became larger. Larger helicopters therefore became more difficult to develop and produce. The mistake of depending on excessively large helicopters became apparent in 1956 when one H-16 prototype crashed.[5] The Army had to cancel this heavy transport helicopter program because of its cost and hopeless mechanical complexity. The XH-17 prototype heavy lift helicopter also never went beyond research and development. Despite the Army's interest in flying crane designs, they were costly, mechanically complex, and had logistical support problems. They therefore had a low funding priority, with most of the research and development money spent on the OV-1 and engine development work.[6]

The H-37 medium transport helicopter's development also ran into trouble and suffered mechanical difficulties.[7] Furthermore, the requirements of all three services far exceeded the limited production volume available. Sharing the H-37 production with the Marine Corps, the Army obtained only 48 by 30 June 1958, and a total of 90 by May 1960 when procurement ended. Even the light cargo helicopter suffered from more demand than supply. During the Korean war and immediately thereafter, Air Force procurement of the H-21 seriously reduced deliveries to the Army. It had to rely on the H-19 utility helicopter as a substitute.[8] Production delivery problems were still present even in FY 1956, though for other reasons. This, together with shortages of maintenance personnel, forced the Army to revise the activation schedule for transport helicopter companies.[9] The Army also modified its force structure plans to take into account the lack of heavy transport helicopters and the small number of medium transport helicopters. The Army made a deliberate policy switch to fixed wing STOL transports and smaller helicopters.[10]

High Initial Procurement Cost

The problem of high initial procurement cost had four causes. First, the need to reduce aircraft weight to obtain more payload was an all-important consideration. To reduce weight, manufacturers used special materials for components and structures. A simple bolt, for example, might cost $1.50, while the same bolt for a tank or truck might cost 10 or 12 cents. A 200 horsepower engine costing $200 to $300 for a tank or truck would cost several thousand dollars for an aircraft. The use of a gas turbine engine with even less weight might cost $30,000 to $40,000. A second cause of the transport helicopter's high initial cost was the lack of competition because of the nature of the helicopter industry.[11] A third reason was the lack of mass production techniques. Essentially, all aircraft were hand-made.[12] Finally, to increase the payload of helicopters beyond a certain size, their technical complexity, and hence cost, would increase by a greater margin. For example, in 1956 the cost of an H-21 or H-34 ($1\frac{1}{2}$–ton payload) was around $229,000, while in 1957 an H-37 (3-ton payload) cost over a million.[13] Unit cost quadrupled as payload doubled. This came under criticism as early as October 1956.[14]

Excessive Maintenance

The problem of excessive maintenance had several related causes. First, the helicopter has to bear all aerodynamic loads on its rotors and other dynamic components, unlike an airplane that bears all aerodynamic loads on its structures. This meant the helicopter's dynamic components wore out faster than the airplane's. These components, known as "time change components," needed replacing after a certain period in operation, before failure occurred.[15] The helicopter's time change components were not only very expensive, they also had a comparatively short life-span that required frequent changes. For example, the main gear box on the H-37 had a cost of

$37,249, but had an original design life of only 150 hours.[16] Since helicopters were also very complex pieces of machinery and did not have design features for easy maintenance, changing these components required many man-hours. It also required comparatively higher skills. Compared with the L-19, the H-21 or H-34 took five times the effort to maintain.[17] A shortage of skilled maintenance personnel further compounded the already low helicopter availability rate. The lack of standardization between different helicopters also complicated the problem of training maintenance personnel.[18]

Spare Part Problems

The problem with spare parts was twofold: cost and parts supply shortages. Shortages of spare parts might occur because of an unanticipated requirement. It might also occur because part usage was higher than anticipated. This usually resulted in a long waiting period when the helicopter was not serviceable. The high cost of spare parts occurred because the Army deliberately overbought certain parts to hedge against unforeseen circumstances or greater usage. For example, the Army bought as many as 100 percent extra transport helicopter transmissions, and 50 percent extra light aircraft engines. This allowed a stockpile from which to draw spare parts when components were due for overhaul. Another cost was the overprocurement of items, in anticipation of a need greater than was actually the case. Also, obsolescence occurred within the large number of stockpiled spare parts. Parts might change either to secure a slightly higher performance, to reduce still further the empty weight, or to correct a deficiency. It was not uncommon for the Army to void 30 percent of parts procured for a particular aircraft. The large numbers of different aircraft and engines required a large infrastructure to control parts and fuel delivery.[19] Brigadier General von Kann summed up the net effects these deficiencies had on the Army's acceptance of air mobility:

> If this air-mobility is so important, why are some commanders approaching it with dignified acquiescence rather than enthusiastic co-operation? Recent surveys point to maintenance worries as the primary cause of this hesitation. There is no doubt that the maintenance requirement coupled with a spare parts shortage sometimes makes the current helicopter a liability as well as an asset—but this

situation is becoming less grim. A strong Army-wide program, now under way, is designed to increase helicopter utilization by new maintenance procedures and more responsive supply back-up.

We must not allow the shadow of the maintenance problems to completely darken our outlook. The helicopter was proven essential in Korea as a relative infant. It still has growing pains, but it is maturing rapidly in its effectiveness. Let's not wait for perfection before acceptance.[20]

Solutions

The Army embarked on a number of general and specific solutions for these problems. One example of a specific solution to a particular problem was the use of tighter parts control mechanisms to improve parts supply and reduce stockpiles.[21] Another example was the use of the Navy and Marine Corps as an alternative procurement agency to the Air Force. This meant the Army had a larger pool of aircraft to choose from, and alleviated some of the production bottleneck. Navy or Marine Corps aircraft procured included the H-19, H-23, H-25, H-34, H-37, and OV-1.[22] An example of a general solution was the improvement of dialogue between the Army and industry. This led to improved maintenance features and procedures, and the design of more durable components. The Army also reorganized its aviation support organizations to provide more efficiency.[23] The main answers for overcoming transport helicopter deficiencies, however, came from three general solutions. They were: the use of comparable fixed wing airplanes as substitutes, accelerated service testing of existing helicopters, and research and development into the next generation of transport helicopters. The accelerated and intensified testing of helicopters under realistic service conditions gave an accurate picture of spare part requirements. It also tested the durability of major components beyond the design life-span. The Army found it could extend considerably the hours of operations of many time change components without endangering safety.[24] This testing enabled components to begin reaching the 1,000-hour life-span the Army sought.[25]

The use of fixed wing airplanes in the same payload class had numerous advantages in comparison with helicopters. Production was relatively abundant, initial costs were reasonable, running costs were lower, and availability higher.[26] The implication of this substitution policy was to change the aviation force structure. It led to the fielding of the U-1 Otter as a substitute for helicopters in the $1^1/_2$–ton payload class. This airplane fell just within the 5,000-pound empty weight restrictions and had a STOL capability superior to that of the L-19, L-20, or L-23.[27] It also led to the controversial AC-1 Caribou STOL transport procurement, which was in the 3-ton payload class.[28] It was the fixed wing equivalent of the H-37. The 1956 Army Aviation Plan and subsequent guidelines stressed the complementary nature of fixed wing and helicopter transports.

> The ability of the helicopter to fly contour and land in minimum air space makes it invaluable for short hauls in the forward areas. Fixed wing transports provide more efficient operation, cost less and can fly faster over longer ranges. However, airplanes require comparatively more airspace for maneuver and landing and must fly higher. Because of these factors, airplanes are more suitable for type missions which are less subject to enemy air or ground action. The aim should be to develop fixed wing aircraft with a vastly improved capability of operating from unimproved landing areas. Successful accomplishment of this objective would permit the fixed wing aircraft to perform a high percentage of the missions now contemplated for the cargo helicopter, freeing the latter for utilization in forward areas.[29]

Research and development into the next generation of transport helicopters was arguably the most important solution. The Army shifted its research and development policy away from larger transport helicopters after the H-16 cancellation. It concentrated on the XH-40 (later redesignated as the HU-1 Iroquois, but commonly called Huey).[30] The HU-1A was originally a utility helicopter designed for the aerial ambulance role to replace the H-19 and H-25. It had a crew of two and could carry two litter patients and one medical attendant, or four walking-wounded. Later, with the development of the 8-passenger B-model and the 12-passenger D-model, it also took on the general transport role. It was to replace the L-20, H-19, H-34, and H-21.[31] It was also important in the Army's armed helicopter plans. It combined in a single airframe the necessary agility and payload-carrying capacity that others lacked. The secret of its success was in its use of the T53 turbine engine and the emphasis on ease of maintenance. Three different T53 variants powered the A, B, and D-model HU-1 (with 700, 1,000, and 1,100 shaft horse power respectively). Turbine engines had less moving parts and vibration, and they gave more power for less weight than reciprocating engines. The T53 was therefore more reliable and required less maintenance.[32] Its smaller size also helped reduce the helicopter silhouette. It was possible to deploy the compact and easily concealed XH-40 with front line troops.[33]

Near the end of the decade, the Army also began to develop the HC-1B Chinook as a replacement for the H-21, H-34, and H-37. This helicopter used turbine engines as well (T55). It was capable of carrying 6,000 pounds of payload for 100 nautical miles and half this load on the return flight. These new developments had the corollary of reducing the number of aircraft models in Army service. In 1961 the Army outlined a plan to consolidate the ten main aircraft types in service to six. The older types included the L-19, L-20, L-23, U-1, H-13, H-23, H-19, H-21, H-34, and H-37. The new types were the HU-1, HC-1, LOH (Light Observation Helicopter), OV-1, AC-1, and L-23F. Fewer models meant fewer spare parts. Also the conversion to turbine power and standardization of fuels to only two or three grades would bring further savings.[34] The more easily maintained turbine powered aircraft entering Army service in the early sixties led to large increases in the aviation force structure, especially within the divisions (see Chapter 7). The goal was to have a total inventory of 8,000 aircraft in 1970, an increase from around 5,500 in 1961.

PROBLEMS IN TRAINING AND INDOCTRINATION

Training and Indoctrination Problems

Although helicopter-borne troops did not need special training as paratroops did, helicopter assault operations were nevertheless much more complicated to plan than ground attacks.[35] Exercise Lift, conducted during May 1954 and involving an infantry battalion, produced many negative appraisals. The 1957 Infantry Instructors' Conference readily recognized that some people in the Army were paying only lip service to air mobility training.[36] After leaving the Army Aviation Center in 1957, General Hutton became the Commanding General of the 8th Infantry Division Artillery in Europe. From there he observed that the Army's air mobility training was in a poor state. This drove him to write an article in October 1958 reemphasizing the first principles of helicopter-borne operations.[37] Another Army aviation officer, Colonel Robert Williams, said in 1958 that the Army did not need to educate or convince field units on the growing importance of Army aviation. He said the issue was really about how to integrate the capabilities of Army aviation into field units.[38] This training and practice problem did not improve even by 1959, and could not match the achievement of the Marine Corps.[39] The Director of Army Aviation, Brigadier General Clifton von Kann, said in May 1959: "Today, battlefield mobility is a prime Army objective. I do not believe we have yet succeeded in exploiting the mobility which Army Aviation has already made available . . . or, as a consequence, in clearly recognizing its future potential."[40]

It was for this reason that the Army published Training Circular 20-1 on 1 May 1959, calling for Army-wide air mobility training for all units. The circular stated that one of the training objectives was to indoctrinate Army personnel with the principle that air mobile operations were a normal part of land combat operations. Moreover, the circular called for units to integrate air mobility training into the current unit training programs. This was to develop the capability of conducting these operations at each level from squad to combat group or regiment (implying that the Army had not been doing so before).[41]

Even after the appearance of this training circular the situation did not improve. For example, in 1960 the battle groups within the 82nd Airborne Division were practicing minor missions with a few helicopters. They used helicopters for logistical support and repositioning the mortar platoon, and as airborne command posts.[42] There was no indication that the division's battle groups had yet managed to practice large air mobile operations. Indoctrination problems were still present during Exercise SWIFT STRIKE in August 1961.[43] The training problem also extended to logistical operations, as Major General Richard Meyer pointed out in April 1962.[44]

Indeed, it was questionable whether the Army could execute widespread air mobile operations even if pressed. For example, air mobile operations needed pathfinder units and equipment for terminal guidance. In 1954 the Chief of Army Field Forces ordered the activation of two such teams. In addition, it planned to have one team for each aviation battalion in the field army. However, even by 1960 the Army only had the original two pathfinder teams.[45] This made a mockery of the view taken in Training Circular 20-1 that air mobile operations were "a normal part of land combat operations."

There were several factors contributing to poor training and indoctrination. One culprit was perhaps the extremely long time it took to complete the key manual, FM 57-35 *Army Transport Aviation—Combat Operations*. The Infantry School started it in May

1955 and completed a draft in June 1955. While awaiting coordination and approval, the Infantry School published this draft as Special Text 57-35-2 in January 1957 for training purposes only. In June 1957 the Department of the Army approved its publication, pending minor revisions.[46] The final version did not appear until June 1958, three years after CONARC initiated work on it. In October 1958 the Army was still in the process of printing the manual.[47] Even when it became available, it would take more time before officers became familiar with it.

Another factor was the existence of indifference towards air mobility. Commanders of mid-size units were especially reluctant to utilize Army aviation.[48] Brigadier General von Kann said in June 1959:

> [M]any of our commanders do not understand aviation and we all tend to recoil from things we don't understand. We must remember that only a few years ago a lieutenant colonel didn't even dare to ask for a helicopter ride. He could not foresee the possibility of having full use of such a vehicle. Now five years later our lieutenant colonel comes back to troop duty a grade higher and in command of a battle group. He now has his own helicopter and control over several other airplanes. Is it any wonder that he is not prepared to utilize these and many more airships in the accomplishment of his mission?[49]

A further factor compounding the indoctrination problem was the basic flaw in the training system. While senior officers attending the Command and General Staff College had instruction on air mobile operations, there was no indoctrination of lower-level officers, staff, troops, and units. The problem was that units, especially the division, had considerable autonomy in their own training.[50] Units could ignore air mobility training if they wanted to. Rather than ordering units to practice air mobile operations, the Army had to "urge" them to do so. For example, in an April 1960 article von Kann urged all unit commanders to plan for company size or even platoon size operations on every possible occasion to realize the transport helicopter's full potential.[51] This suggests that the Army was not even practicing the smallest air mobile operations, let alone battle group size operations. An added difficulty, von Kann said, was that training people to use aviation effectively did not come easily, especially when flying in and out of dirt strips or confined areas. He noted that combat maneuvers in these places were frequently dangerous and always time consuming to set up and rehearse. Moreover, he added:

> [B]ear in mind that it takes time to train aviators to work with tactical units; and an even more critical factor is the time it takes the tactical units to work with aviation. This is not a serious problem in flying from airfield to airfield, but when the destination or pickup point is a small clearing in the woods, it becomes a problem of different dimensions. . . . [I]t takes approximately a year to develop a truly professional standard of teamwork between the infantry units of a division and the aviation which must work with these units.[52]

Furthermore, conflicting flying requirements made the development of close teamwork very difficult. In the late fifties there was an emphasis on instrument qualifications and safe flight operations that was incompatible with tactical flying.[53] To acquire and maintain instrument qualification took much work and practice. As a result, support of ground troops suffered. An aviation staff officer, Major John Stockton, pointed out in 1958 that Army aviators were to blame for neglecting tactical flying. He said that once aviators had become instrument-qualified, they began considering themselves "miniature airline captains." Stockton added that they were "all of a sudden above the dirty business of flying in and out of cow pastures, especially at night." They were also "no longer interested in low level navigation flying procedures or in the difficult art of acquiring information for our ground brethren." He said that they began to prefer flying in established air corridors from one fixed airfield to another at airline or Air Force altitudes.[54] Peacetime safety requirements were partly to blame for this mentality. Army aircraft had to remain higher than 150 feet during training.[55] The emphasis on safety was due to the comparatively high Army aviation accident rates. In fiscal year 1958, accidents involving Army aircraft caused more than 40 percent of the total dollar loss due to all accidents in the Army. The cost of Army aircraft accidents was more than $9,500,000. This compared, for example, to a fire loss of approximately $5,000,000.[56]

Perhaps the most important factor contributing to poor air mobility training was the insufficient number of transport helicopters. General Howze stated in 1956 that the Army planned to utilize helicopter mobility on a large scale. However, he said the problem lay with the

Table 5-1
Total (CONUS) Activated Aviation Units, 1955–1960

Date	Battalion HQ	Light Helicopter Co	Medium Helicopter Co	Fixed Wing Transport Co
1 July 1955	3 (3)	10 (7)	–	1 (1)
1 July 1956	5 (4)	14 (9)	–	2 (2)
30 June 1957	6	18	–	3
1 July 1958	6 (4)	19 (11)	2 (2)	4 (2)
1960	6 (4)	21 (10)	4 (3)	5 (4)

inadequate number of transport helicopters. "Helicopters only now [1956] are beginning to arrive in sufficient numbers to be a major factor in troop movements. It will take a large number of machines to constitute a real factor, but we'll get them."[57] Moreover, the need to send transport aviation units overseas to large Army contingents such as the United States Army in Europe compounded the problem (see Table 5-1).[58] Even within the United States itself, the Army had to disperse transport aviation units to the various Army elements around the country.[59] Each light ($1\frac{1}{2}$–ton) transport helicopter company could only lift one infantry company at a time. A division training to perform attack or counterattack in a nuclear setting would need to find several such companies to lift entire battle groups. This would be very difficult to arrange.

Other factors further aggravated the shortages. There was a delivery problem and a shortage of maintenance personnel, which affected the activation schedule of helicopter companies. The shortfall applied in particular to the H-34. This led the Army to cancel one training class in 1956. Moreover, the Army needed more aircraft for the ROTAD test (101st Airborne Division in Exercise JUMP LIGHT) in November 1956 and the SKY CAV II training commencing in January 1957. (Exercise SLEDGE HAMMER incorporating the SKY CAV II test occurred in May 1957.) Furthermore, the new Pentomic division structure involved a significant increase in utility and light transport helicopters. It was necessary to divert personnel and equipment to fulfill these two requirements, further aggravating the activation schedule of transport aviation units.[60]

Lack of Dedicated Air Mobile Units

Brigadier General Carl Hutton argued in 1955 that the solution to the indoctrination problem was to make the Army air-minded in the next ten years. He said that thirty years earlier almost everyone in the Army was horse-minded. In 1955 everyone was motor-minded. Soldiers, Hutton said, had to acquire air-mindedness:

> The basic ingredient is pilot qualification, just as equitation is the foundation for horsemastership and the ability to drive a car is the foundation for motor-mindedness. A great deal of aviation lore can be learned from books and from secondhand sources, but sooner or later the air-minded soldier is going to arrive at a point where only learning to fly will satisfy his need for additional knowledge. There is no other way for him to gain confidence in his own ability to understand the problems of aviation, or in his capacity to make sound decisions.[61]

In essence, Hutton was arguing for making the ability to fly as widespread and as normal as the ability to drive a jeep or truck. However, this was far from practical, as flying was an order of magnitude more difficult than driving. Obtaining a reasonable proficiency in flying took much time and effort, a problem not easy to solve, as Hutton acknowledged in 1957.[62] Moreover, the quality of personnel required for flight training was comparatively high, as indicated by the stringent selection process and high "washout" rates.[63]

Dedicated air mobile units (ground combat units with integral transport aviation) would be a more logical solution to overcome the training problem. These units would have an added advantage of being optimized for helicopter assault operations. There was no shortage of such a proposal, especially after Gavin's 1954 article on sky cavalry. Army staff members within the Pentagon had visions of future three-dimensional warfare.[64] Hutton publicly voiced the need for large-scale air mobile units.[65] Colonel Jay Vanderpool at the Combat Development Office, Fort Rucker, drew up one of the earliest completely air-mounted sky cavalry organization. This became a building block for his later proposal, the ARMAIR Brigade. The brigade consisted of some 4,000 men and 630 air vehicles.[66] Even the 1958 Army Aviation Guidelines noted this study.[67] General Howze was well aware of the ideas propounded by Hutton and Vanderpool.[68] In late 1957 Howze himself proposed forming an air cavalry battalion with armed helicopters to the Army. CONARC took this up and became the proposing agency for the idea.[69] This proposal even reached the Army Chief of Staff in 1958.

The Army did not establish any dedicated air mobile units for practical and operational reasons, rather than due to conservatism. The lack of transport helicopter resources was the practical reason mitigating against dedicated air mobile units. The Army's operational priority on the nuclear battlefield was another. The need to make all ground units more mobile was an overriding concern for the Army high command. Anything that detracted from this would not interest it. Without enough helicopters to go around, a helicopter-intensive unit was out of the question. It would deprive the needed mobility from the rest of the Army. Moreover, the Army may be reluctant to create a new type of specialized unit similar to an airborne unit while there was already a shortage of main line units. Therefore, although the proposal from Howze for an air mobile battalion reached the Chief of Staff, the Army deliberately rejected the idea in favor of making ordinary units air mobile.[70]

The Army and the Infantry School reflected this priority by emphasizing the ability of ROTAD and ROCID units to execute air mobile operations, and trying to give these units such additional training.[71] Another official Army policy outlined in the Combat Development Objectives Guide (CDOG) reinforced the "air mobility for all" policy:

> As technological advances permit, aerial vehicles will be assigned in increasing numbers to lower echelons of command. During the interim period, aerial vehicles will be assigned to headquarters one or two echelons above those subordinate units having an intermittent requirement for their use. When a unit has a continuing requirement for aerial vehicles, they will be made an organic part of the unit.[72]

While the Aviation Center had an interest in dedicated air mobile units, it had to listen to other segments of the Army in air mobility development. This was because there was no aviation branch, so it was a "servant" of other user branches. The Combat Developments Office at the Aviation School had to mirror this policy to integrate aircraft within the division's subordinate units. One aviation officer began to question this organic tendency and insinuated that it stunted combat aviation development.[73] However, this was not necessarily so. The Infantry School accepted the idea of self-contained air mobile units of company size. Components of this unit would include armed helicopters for direct fire, aerial artillery, and air mobile infantry.[74] The Aviation School was also working on the organization and tactics of various air mobile units of company size. Moreover, it also believed that other aviation organizations besides the integrated units would require the formation of an aviation battalion at the division level.[75]

An alternative to creating new air mobile units was to convert airborne units into helicopter-borne forces. Some elements in the Army considered this very early on. Some even considered using standard infantry divisions for helicopter operations, thus obviating the need for specialized airborne divisions.[76] This was a possibility because of the similar goals of airborne and helicopter assault operations. Compared to the helicopter, the parachute delivery of men and equipment was relatively inefficient (see Table 5-2). Moreover, specialized equipment at $400 per paratroop was expensive, not to mention the specialized training required.[77] Parachutes were also rather heavy and presented an overhead that troop carrier aircraft had to carry. Once on the ground, it was difficult to extricate parachute forces. This might present some problems with the vulnerability of such forces to nuclear weapons. Also, it took a parachute force 20 to 30 minutes to assemble after a jump, while helicopters could land a force in whatever tactical formation desired.[78] Therefore the Army had a continuing desire to replace the parachute during the Eisenhower era.

The airborne community itself saw airborne operations as shackled to "an immense concrete anchor."[79] The need for airfields and the slow marshalling process presented vulnerability problems. It also realized the helicopter's important role in airborne operations as a result of the rapid development of antiaircraft guided missiles. It appreciated the difficulty of executing a conventional mass airborne operation when hostile area air defenses were in place. A 1957 airborne conference concluded: "Small rotary winged aircraft formations can now penetrate missile defended areas and will remain capable of so doing in smaller and smaller groups for the foreseeable future."[80] The conference felt that conventional parachute assaults were obsolete under intense air defenses. Single file helicopter formations penetrating underneath the air defense envelope were the only viable option.[81] One observer at this 1957 Airborne Conference noted that there was a very obvious "romance" between the airborne and Army aviation communities.[82] It was therefore not surprising that airborne officers recommended the procurement of helicopters as fast as production allowed.

However, helicopters had an inherent range limitation, with no possibility that future development would make them comparable to fixed wing aircraft. This meant that the Army still had to use troop

Table 5-2
Relative Advantages and Disadvantages of Helicopter Versus Parachute/Troop Carrier Operations

Helicopter Advantages:	Parachute/Troop Carrier Advantages:
• Does not require large airfields for departures and arrival. • More conducive to maintaining secrecy. • Less affected by weather conditions than fixed wing aircraft. • Loading areas less vulnerable to attack. • Helps preserve tactical integrity of forces. • Can lift awkward size and shape of equipment in sling-loads. • Can keep forces mobile at objective. • No special training required. • Quicker rate and ease of operation can take advantage of local or fleeting situation, for raids and dealing with hostile airborne raids. • Can extricate assault forces.	• Can prepare for operations far from objective and fly there rapidly. • Gives intertheater and intratheater mobility. • Such units in theater reserve pose constant strategic threat to enemy.
Helicopter Disadvantages:	**Parachute/Troop Carrier Disadvantages:**
• Cannot operate over long distances (over 100-mile range or 50-mile radius). • Lifting capability reduced by hot weather and high altitudes. • High costs and limited availability of helicopters.	• High cost of training and equipment. • Three weeks of specialized training. • Marshalling area more vulnerable to attack. • More difficult to conceal operations. • More susceptible to rain, low cloud, and high winds. • Need time to reorganize dispersed units after parachute descent. • More vulnerable to antiaircraft defenses. • Mobility severely reduced at objective area. • Loss of men and equipment from parachute descent.

carrier aircraft and the parachute for long-range strategic operations.[83] The retention of a parachute capability was necessary in case the destination had no suitable airfields. The airborne community, realizing that parachutes would remain the chief means of airborne delivery, envisaged using the helicopter in conjunction with assault transport aircraft rather than replacing it. Helicopters were more useful for short-range or small-scale operations.[84] The only comparison possible between helicopter and strategic airborne operations was if the helicopter force used amphibious assault ships to give it strategic reach.[85]

Cost relative to payload was also a factor. For example, in 1956 the average cost of an H-21C or H-34 helicopter was $229,000, while one C-119 fixed wing transport cost $375,000.[86] The C-119 could carry 16,000 pounds, while the H-21 and H-34 could only carry 2,664 and 2,512 pounds respectively for a 100-nautical-mile combat radius. However, one cannot directly compare helicopters and troop carrier aircraft. They each have their advantages and disadvantages depending on the operational need.

The airborne community realized that convertiplanes offered the versatility of the helicopter while offering the speed and range of conventional airplanes. Therefore the Army continued to push convertiplane research, hoping ultimately to replace the assault transport, itself a replacement for the glider.[87] It did realize, though, that there were still many years before a transport convertiplane would be available. In the meantime an improved short take-off and landing aircraft was to act as an interim solution.[88]

Despite the increased reliance on helicopters, the Army did not convert airborne units wholesale to air mobile units. It was not just the shortage of transport helicopters. A more important reason was the transport helicopter controversy between the Army and the Air Force in 1955–1956. The main point of contention was the movement of Army troops (both regular and airborne) by helicopters. The Air Force contended that all such missions were "airborne" operations and therefore should use Air Force helicopters. The Army, which wrestled control of transport helicopter operations from the Air Force, claimed they were not airborne operations at all.[89] Hence the Army could not use helicopters in any way that could let the Air Force claim its case. This may explain the Army's reluctance to convert airborne units to air mobile units with their own organic transport.

SKY CAVALRY DEVELOPMENTAL PROBLEMS

After the appearance of Gavin's article, the Department of the Army was very much supportive of sky cavalry development. The organization for such a concept underwent a process of metamorphosis until 1961. At first, the Army envisaged mixing aerial reconnaissance elements with ground reconnaissance elements in a single team. This team would also contain electronic and photographic surveillance devices for use both in the air and on the ground. Fixed wing aircraft and helicopters would be joined with tanks, jeeps, and armored personnel carriers in a unit organized and equipped for performing cavalry missions. Integration of the air and ground element occurred at the company level. Its intended purpose was to improve the reconnaissance capabilities of the divisional ground cavalry unit.[90] From 1955 to 1957 the Army tested three different experimental sky cavalry organizations (for supporting different types of division) during three major exercises (SAGE BRUSH,[91] JUMP LIGHT,[92] and SLEDGE HAMMER).[93] Because of the infancy of the concept and many teething problems, the earlier test reports were not encouraging. The difficulty was in integrating air and ground elements into a coherent organization. If it had not been for the support from higher headquarters, the survival of the sky cavalry concept was doubtful and it would probably have ended prematurely.

For example, during Exercise SAGE BRUSH, the sky cavalry test received the personal attention of the Army Secretary and the Chief of Staff. Initially, sky cavalry was a contentious issue with the Air Force. Army Secretary Brucker had to wrestle from the Air Force the right to operate sky cavalry on an experimental basis for the duration of the exercise.[94] When General Taylor went down to the Louisiana exercise area, the sky cavalry troop was the first unit he visited despite it being off the beaten track. He also publicly announced in a press conference that he had a keen interest in sky cavalry. He said the Army had a dire need for this kind of organization that could sweep behind enemy lines, gather information, and spot targets. In addition, Taylor was eager to obtain the evaluation report on sky cavalry. He pressed for the answers and military organization recommendations at the earliest possible moment.[95] This suggests that the Army had high hopes for implementing sky cavalry as soon as possible, not least because Sky Cav was a catchy term with mass appeal.[96] Even discouraging test results did not put the Army off the sky cavalry implementation plans.[97]

As a result of deficiencies in sky cavalry organization and its tactical concept, the Pentomic reorganization that occurred from November 1956 did not include a full divisional sky cavalry unit. However, it did incorporate a limited implementation with the target acquisition section under the direct support platoon.[98] This section was for direct support of the reconnaissance and surveillance platoon of the divisional cavalry squadron. In addition, the limited number of utility helicopters (and cargo helicopters in the armored division) could lift elements of the reconnaissance battalion on sky cavalry missions.[99] The reason the Army persisted with the sky cavalry concept despite earlier setbacks was that it recognized the need for such an organization on the dispersed nuclear battlefield. It also saw sky cavalry as a new military development in its own right, and was conscious of its importance despite its infancy.

Meanwhile, from 1956 the Aviation Center initiated the development of a completely air-mounted version of sky cavalry. This became known as the Aerial Combat Reconnaissance (ACR) company or air cavalry.[100] Rather than using aircraft as a complement or supplement to ground elements, aircraft were the primary vehicles of this concept. This air fighting organization was capable of executing independent tactical tasks.[101] Key officers involved at the Aviation Center envisaged completely air fighting units of brigade and division size, with their ACR company as the basic building block. It also offered a rationale and impetus for the development of armed helicopters.[102] However, using helicopters as a weapons platform was a relatively unexplored area, with little basic information available.

The development of armed helicopters and the ACR concept and organization at first only had the tacit approval of CONARC.[103] The main center of activities was at the Army Aviation Center,[104] with participation and some activities at the Infantry and Armor schools.[105] The experimental unit promoted the ACR concept by tactical and firepower demonstrations during Army conference gatherings. Top officials, including Taylor and even Eisenhower, were witness to these. It received the Army's approval and expanded to company size because it showed more promise than the combined air and ground version. As the failure of the integrated type became apparent after 1957, the Department of the Army eventually adopted the ACR company as the basis for its sky cavalry.[106] By late 1957, Secretary Brucker publicly endorsed the idea of armed helicopters in sky cavalry units.[107] Brigadier General Carl Hutton, who started up the ACR unit, said:

> This concept of sky cavalry has been the easiest to sell itself. Experiments are being conducted using armed reconnaissance units. Of course, in my opinion, this testing is really for the purpose of developing tactics, technique, and organizational details because I consider sky cavalry to be so eminently sensible that it is inevitable. We only need think of the value we could have received from such units during the Korean War in sealing off the Yalu River to see what they will be worth. Sky cavalry units would have meant the difference between stalemate and victory if they had been technically feasible in 1950. They are technically feasible now.[108]

In December 1958 the Department of the Army decided to field test an ACR-type organization. The test began in 1960. The Army gave the organization the name Aerial Reconnaissance and Security Troop (ARST) to distinguish it from previous sky cavalry units. The report on this test was favorable towards the concept.[109] Moreover, by the end of the decade, a combined-arms, completely air mobile cavalry unit with a mobility differential had gained wide currency within the Army. The inclusion of air cavalry units in the MOMAR study was evidence of this acceptance. Even Lemnitzer reported that the ARST tests at Fort Stewart were promising.[110]

The Rogers Board deliberations in 1960 presented an opportunity for forming a large air cavalry unit. While having a predominantly research, development and procurement orientation, aviation proponents used the board as a vehicle to propound radical concepts. General Howze, a member of the board, submitted a requirement for air fighting units in the last section of the Rogers Board report. (The report attached the Howze memorandum as an enclosure because it went beyond the charter of the board.) He called for the establishment of five air cavalry regiments with 175 utility and 85 light observation helicopters each. Each active corps would have one such regiment. He stated the rationale for these units:

> In the days when the horse provided the highest degree of battlefield mobility, it would have been a fundamental error to restrict the assignment of horses to the infantry divisions. While infantry divisions employed horses in considerable quantities, with benefit, it was necessary and desirable to group a substantial percentage of all the horses in cavalry units in order to take proper advantage of their mobility.[111]

He acknowledged that the assignment of aircraft to modern Army units would improve their ability to execute their conventional missions. However, this would limit the aircraft's mission capability to that of wheeled mobility. He argued that it would be necessary to devote some aircraft to form air cavalry units to exploit air mobility to the full. He estimated that the total cost would be approximately equivalent to one Polaris nuclear submarine. If the air cavalry replaced other units in the force structure, the savings would partially offset the costs involved. He said that air cavalry would find particular applicability in any battle area in which the threat of area weapons forced wide troop dispersion. It would also be applicable in "brush fire" actions against relatively unsophisticated opponents. The Rogers Board agreed that its procurement objectives were in line with Howze's proposals. Moreover, there were sufficient aircraft in the existing inventory to activate an experimental unit.[112]

When Lemnitzer approved the board's recommendation on 19 March 1960, he also charged the Deputy Chief of Staff for Operations (DCSOPS) with the responsibility for preparing recommendations on air fighting units.[113] Unfortunately, the ARST report from CONARC, submitted in August 1960, stated that there was no requirement to test a battalion size unit. This probably reduced the possibility of the Army forming a regimental size unit. Moreover, events associated with a new presidential administration soon overtook the actions of DCSOPS. Nevertheless, the Army came close to achieving a breakthrough in air mobility in 1960. Lemnitzer's action was a good indication of his readiness to accept the radical idea.

By 1958 the Army had also determined in principle the practicality of armed helicopters, although confined to crude modifications and riggings of armament kits.[114] This was important for future air mobility developments. It was easy to use armed helicopters developed for air cavalry missions to act as fire support for air mobile operations. The Army did not miss the next window of opportunity in 1961, when another reorganization occurred, to establish regular air cavalry units and assign armament kits to helicopters (see Chapter 7).

NOTES

1 (1) Pack, p. 2. (2) See also MRRP, "Review of Army Helicopter Program," Staff Study 13 and 14.

2 "Army Will Buy 1000 Cargo 'Copters," *AA*, 29 March 1954, p. 21.

3 Colonel Bunker was Chief of Air Transportation Services Division in 1953, and Assistant Chief of Transportation for Army Aviation in 1954. When promoted to Brigadier General he became Commander, Transportation Supply and Maintenance Command, in 1958.

4 (1) Harry S. Baer, Jr, "Army Spells Out Future Helicopter Needs," *AA*, 24 May 1954, p. 13. (2) "Howze Asks Improvement in 'Copters," *AT*, 12 May 1956, pp. 1, 10.

5 (1) House of Representatives, *Army Appropriations FY 57*, pp. 334–35. (2) Weinert, p. 156. (3) "YH-16 Cancelled," *AW*, 23 April 1956, p. 38.

6 (1) "Army Awards Contracts for 'Flying Crane' 'Copter," *ANAFJ*, 18 August 1956, p. 1592. (2) "Army Lets Five Study Contracts to Probe Flying Crane Capability," *AW*, 20 August 1956, p. 30. (3) "Flying Crane Ideas Submitted to Army," *AW*, 11 November 1957, p. 64.

[7] "Helicopter Design Lags Behind Needs," *AW*, 12 March 1956, pp. 259, 265.

[8] (1) TC, "Army A/c P&P Programs," pp. 41–4. (2) Bristol, "Oral History," tape 1, pp. 22–23.

[9] Weinert, pp. 197–202.

[10] DA, "1956 AAP," Annex 2, Appendix A, p. 2.

[11] Brigadier General William B. Bunker, Commander, Transportation Supply and Maintenance Command, "Supply Management Problems in Aircraft Logistics," *USAAD*, April 1957, pp. 10–12.

[12] Brigadier General Carl I. Hutton, Commanding General, Army Aviation Center, "Universal Superhighway," *USAAD*, May 1956, p. 4.

[13] (1) House of Representatives, *Army Appropriations FY 57*, p. 332. (2) DA, "1957 AAG," Annex 2, Appendix A, Inclusion 1.

[14] (1) "Fixed-Wing Concepts Block True Helicopter Future, General Says," *AW*, 15 October 1956, p. 33. (2) Brigadier General Carl I. Hutton, Commanding General, Army Aviation Center, "Economics of the Military Helicopter," *USAAD*, October 1956, pp. 3–4.

[15] Bunker, "Supply Management Problems," pp. 7–9, 12–13.

[16] (1) J. S. Butz, Jr, "Army Begins Cost Cutting Program Based on 1,000 Hr. Test Results," *AW*, 9 September 1957, pp. 52, 57, 61. (2) See also: Claude O. Witze, "High Costs Imperil Army Copter Program," *AW*, 17 May 1954, p. 14.

[17] The Army measured maintenance requirements in terms of "aircraft equivalents." An aircraft equivalent was the maintenance equivalent of one L-19 observation airplane. In 1958 the maintenance equivalents of other Army aircraft were: L-20: 1.2; L-23: 2; H-13: 2.5; H-19: 3.5. See (1) Colonel Frank G. Forrest, Aviation Officer, Sixth US Army, "Army Aircraft Maintenance in the Pentomic Army," *USAAD*, September 1958, p. 17. (2) "Present Copters Inadequate: Army," *AW*, 10 May 1954, pp. 17–18.

[18] (1) Lieutenant James H. Muller, Army Aviation Division, TRADCOM, "Army Aviation Division—TRADCOM," *USAAD*, June 1955, pp. 33–36. (2) Larry M. Hewin, Deputy Chief, Army Aviation Division, TRADCOM, "Helicopter Future," *USAAD*, April 1956, p. 12.

[19] (1) Bunker, "Supply Management Problems," pp. 14–16. (2) "Maintenance Crisis Threatens Mushrooming Army Aviation," *AW*, 8 October 1956, p. 28.

[20] von Kann, "Mobility Now," p. 43.

[21] (1) "Support of Aviation Now a Major Mission of Army TC; Management Methods Told," *ANAFJ*, 7 September 1957, p. 26. (2) "Logistics: Lifeblood of Army Aviation," *AA*, 10 March 1958, pp. 56, 58. (3) Lieutenant Colonel William B. Dyer, "Keeping Army Aircraft Aloft," *NDTJ*, March–April 1958, p. 80.

[22] TC, "Army A/c P&P Programs," pp. 40, 42–44, 49, 51–52.

[23] (1) Alexander McSurely, "Army Aim: More Reliable Copters," *AW*, 25 May 1953, p. 20. (2) Department of the Army, Office of the Chief of Transportation, "The Army Aviation Depot System: Its Origins and Development," Washington, DC, 15 October 1959, pp. 1, 4. (3) Harry A.

Jacobs, "Servicing Army Aviation," *NDTJ*, September–October 1953, pp. 66–67.

[24] (1) "How Army Aviation Tests Its Aviation Equipment," *AA*, 10 March 1958, pp. 42, 47, 50. (2) Butz, "Army Begins Cost Cutting Program," pp. 52–54, 57. (3) Dyer, p. 61. (4) "Ft. Rucker Is Heartbeat of Army's Aviation, *AA*, 9 March 1959, pp. 50–51.

[25] Claude O. Witze, "Army Sets Copter Overhaul Target," *AW*, 24 May 1954, p. 18.

[26] (1) Lieutenant Colonel Frank G. Forrest, "Helicopter or Fixed Wing?" *The Army Combat Forces Journal*, June 1955, pp. 46–47. (2) Howze, "Last Three Years," pp. 7–8.

[27] (1) "Army Activates First Fixed Wing Plane Unit," *AW*, 13 June 1955, p. 18. (2) Weinert, pp. 67–68. (3) Infantry School, "1957 Instructors' Conference," p. 40.

[28] Weinert, pp. 16–17.

[29] DA, "1956 AAP," Annex 1, Appendix B, p. 2.

[30] (1) DA, "1957 AAG," Annex 1, Appendix B, p. 8. (2) DA, "1958 AAG," Annex 1, Appendix B, p. 10.

[31] (1) Infantry School, "1956 Instructors' Conference," p. 113. (2) "Army Plans 2,500 Additional Aircraft by '70; Will Drop Types from 15 to 7," *ANAFJ*, 7 May 1960, pp. 1, 16.

[32] (1) Frank Coleman, "Turbine Powered Bell XH-40," *AH*, December 1955, pp. 10, 23. (2) "Bell Helicopter Corporation Announces Two New Army Orders," *AH*, September 1957, p. 13.

[33] "Army's New Aviation Chief Deplores 'Starvation Basis' Support," p. 18.

[34] (1) "Plans Call for Fewer but Better Aircraft," *AA*, 10 March 1958, pp. 35–36. (2) Weinert, pp. 156, 158. (3) Major Donald N. Gower, Project Officer, Doctrine and Organization Division, Combat Developments Office, US Army Aviation School, "Problems Connected with New Air Vehicles," *USAAD*, January 1962, p. 24. (4) "Army Will Add 2500 Aircraft; Modernization Plans Cut Costs," *ANAFJ*, 1 July 1961, p. 8.

[35] (1) Lieutenant Colonel Hubert S. Cunningham, "Air Mobility Staff Planning," *Infantry School Quarterly*, January 1956, pp. 43–46. (2) Captain Joseph O. Wintersteen, Jr, "Helicopterborne Operations," *Infantry* , April 1957, pp. 20–33. (3) Buchanan, p. 7.

[36] (1) Douglas, pp. 20–23. (2) Infantry School, "1957 Instructors' Conference," p. 49.

[37] Hutton, "Helicopter-Borne Operations," p. 5.

[38] "Interview with Army's No. 1 Aviator," *AA*, 10 March 1958, p. 55.

[39] "Letters: Marine on Helicopters," *The Army Combat Forces Journal*, December 1954, p. 8.

[40] "B/G von Kann Will Head Army Aviation," *ANAFJ*, 9 May 1959, p. 7.

[41] (1) Bourjaily, pp. 1, 28. (2) "Airmobility Training Broadened in Army," *AFT*, 8 May 1959, p. 18.

[42] Captains Henry G. Moseley and R. G. Steuart, "Battle Group Aviation Team," *Army*, October 1960, pp. 54–57.

43 Lieutenant Colonel Donato N. Vincent, CO, 3rd Transportation Battalion, "Exercise SWIFT STRIKE," *USAAD*, November 1961, pp. 1–5.

44 Major General Richard D. Meyer, Deputy Chief of Staff for Logistics, Headquarters, USCONARC, "Logistical Employment of Army Aircraft," *USAAD*, April 1962, p. 13.

45 (1) Captain Larry S. Mickel, " The Marines Have Landed . . . ," *Infantry*, February–March 1960, pp. 68, 70–71. (2) Lieutenant Thomas A. Deemer and Captain Larry S. Mickel, "Can the Army Land Too?" *Infantry*, February–March 1960, pp. 69, 72.

46 Infantry School, "1957 Instructors' Conference," pp. 23, 27.

47 Hutton, "Helicopter-Borne Operations," p. 8.

48 See also: "B/G von Kann Sees Army Aviation as Key to Survival," p. 20.

49 "Army's New Aviation Chief Deplores 'Starvation Basis' Support," p. 18.

50 Major General Hamilton H. Howze, Director of Army Aviation, ODCSOPS, "Notes from the Pentagon: Aviation Company Field Training Exercise," *USAAD*, December 1957, pp. 2–3.

51 von Kann, "Mobility Now," p. 35.

52 "Army's New Aviation Chief Deplores 'Starvation Basis' Support," p. 18.

53 "Army Issues New Plea for Relief from Aircraft Size-Weight Limits," *AFT*, 13 June 1959, p. 18.

54 Major John B. Stockton, "Let's Bring Army Aviation on the Team," *Armor*, November–December 1958, p. 31.

55 Moseley and Steuart, p. 56.

56 Brigadier General Ernest F. Easterbrook, Commanding General, US Army Aviation Center, "Army Aviation Philosophy," *USAAD*, October 1959, p. 2.

57 Jules Witcover, "US Army Aviation: Where Is It Going?" *ANAFR*, 6 October 1956, p. 15.

58 "USAREUR Getting More 'Copters," *AT*, 25 February 1956, p. 6.

59 "More Units Joining 5th Army Aviation," *AT*, 3 March 1956, p. 27.

60 Weinert, pp. 197–202.

61 Brigadier General Carl I. Hutton, "The Commandant's Column: Random Shots," *USAAD*, November 1955, p. 3.

62 (1) Hutton, "Without Trumpet and Drums," pp. 141–42. (2) See also: Warrant Officer Harold D. Wilcoxen, Instructor, Department of Rotary Wing Training, US Army Aviation School, "I Flew for Aerial Combat Reconnaissance," *USAAD*, August 1959, pp. 10–11.

63 (1) "Aviation School Pre-Flight Tests May Cut 'Copter Pilot 'Washouts,'" *AT*, 3 March 1956, p. 27. (2) "Army Seeks Helicopter Pilots," *AT*, 24 March 1956, pp. 1, 10.

64 "Army to Fly Over Obstacles by 1970," *AT*, 26 November 1955, p. M6.

65 Bob Williams, "Salute to Fort Rucker," *ANAFR*, 26 November 1955, p. 23.

66 (1) Aviation School, "CDO Packet No. 1," Tab G, ARMAIR Brigade. (2) United States Army Aviation School, "Annual Army Aviation Instructors

Conference, 1959: Report of Proceedings," Fort Rucker, Alabama, 11–12 August 1959, p. 205.

[67] (1) DA, "1958 AAG," Annex 1, Appendix B, p. 4. (2) Weinert, pp. 131–34.

[68] Tolson, *Airmobility*, p. 7.

[69] Weinert, p. 138.

[70] (1) Colonel Jay D. Vanderpool, Director, Combat Developments Office, Army Aviation School, Disposition Form to Assistant Commandant/Commandant, Army Aviation School, "DCSOPS, DA, ACR Test Proposal," 17 October 1958. (2) Colonel Claude Shepard, ODCSOPS, Memorandum to Combat Developments Office, Army Aviation School, "Draft Summary Sheet," n.d. [c.1958].

[71] (1) Infantry School, "1957 Instructors' Conference," pp. 35–38, 49. (2) "The Changing Face of the Infantry," *Infantry*, January 1960, p. 33.

[72] Lieutenant Colonel Morris G. Rawlings, Deputy Director, Combat Developments Office, US Army Aviation School, "The Objectives of Army Aviation," *USAAD*, July 1962, p. 3.

[73] Captain Paul C. Swink, Jr, "How Modern Is Our Concept?" *USAAD*, June 1961, pp. 8–9.

[74] Pinney, p. 65.

[75] Colonel Allen M. Burdett, Jr, Director, Combat Development Office, US Army Aviation School, "Let's Integrate Now!" *USAAD*, February 1961, pp. 8–9.

[76] (1) Lieutenant Ivan I. Slavich, Jr, "Paratroops or Helitroops," *Infantry School Quarterly*, January 1956, pp. 88–93. (2) Lieutenant Colonel John M. Kinzer, Instructor, Command and General Staff College, "Airborne Assault by an Infantry Division," *MR*, October 1953, pp. 45–53.

[77] Slavich, p. 90.

[78] James J. Haggerty, Jr, "No More Paratroops?" *Collier's*, 18 March 1955, p. 25.

[79] Infantry School, "1957 Instructors' Conference," pp. 26–27.

[80] United States Army, "Report of the 1957 Army Airborne Conference," Fort Bragg, North Carolina, 7–10 May 1957, p. 16. Hereafter cited as Army, "1957 Airborne Conference."

[81] Army, "1957 Airborne Conference," Appendix B3, p. 252.

[82] Lieutenant Colonel Robert B. Rigg, "Airborne Warriors Look to the Future," *Army*, September 1957, p. 26.

[83] Kinzer, pp. 49–51.

[84] JATB, "1953 Conference," Appendix A2, pp. 51, 57; Appendix A9, pp. 134–36; Appendix A11, p. 153.

[85] Major Archie J. Clapp, United States Marine Corps, S3, Marine Composite Squadron One, "Their Mission is Mobility," *MR*, August 1953, pp. 14–16.

[86] (1) House of Representatives, *Army Appropriations FY 57*, p. 332. (2) Slavich, p. 90.

[87] The Army and the Air Force originally conceived assault transports as a type of "powered glider" that could take off without another aircraft

dragging it. It eventually developed into fully equipped transport airplanes, with the ability to land and take off from short rough fields.

[88] (1) JATB, "1953 Conference," Appendix A10, pp. 143–44; Appendix C, pp. 268–69, 271–72. (2) Infantry School, "1957 Instructors' Conference," p. 24.

[89] Goodhand, "Army Aviation Policy," pp. 5–6.

[90] Lieutenant Colonel John T. Collier, "Cavalry of the Sky," *AID*, November 1957, pp. 24–25.

[91] (1) DoD, *Semiannual Reports, Jan.–June, 1956*, p. 84. (2) Max Simon, "Cavalry Now Wears Wings as a Reconnaissance Unit," *AT*, 16 March 1957, p. 4. (3) McKenney, pp. 13–18. (4) Howze, "Last Three Years," pp. 14–15, 58–59. (5) Weinert, pp. 119–23. (6) Infantry School, "1956 Instructors' Conference," p. 134.

[92] Army, "1957 Airborne Conference," Appendix A10, pp. 167–72.

[93] (1) United States Army, HQ CONARC, Fort Monroe, Virginia, Memorandum to Commanding Generals Fourth Army, Army Intelligence Center, Chief Signal Officer, Commandants Army Aviation School, The Armor School, The Infantry School, "Instructions for Organization, Testing, and Evaluation of Provisional Sky Cavalry Company (SKY CAV II) (U)," 21 September 1956, Inclosure 1. (2) Weinert, pp. 126–30. (3) Collier, p. 28.

[94] (1) "Copter Controversy," *AW*, 28 November 1955, p. 11. (2) "'Without Prejudice,' Says Air Force," *ANAFJ*, 26 November 1955, p. 18. (3) "Sagebrush Hostilities Involve Army and Air Force Secretaries," *ANAFR*, 26 November 1955, pp. 1, 37. (4) "Armed Services: Dogfight," *Newsweek*, 5 December 1955, p. 23. (5) "The Army's Month: Best Combinations," *The Army Combat Forces Journal*, January 1956, p. 8.

[95] "Army Looks to Sage Brush's SkyCav Operation as Basis of Permanent Unit," *ANAFJ*, 10 December 1955, pp. 1, 3.

[96] William E. Vance, Editorial Staff, "Saber in the Sky," *USAAD*, August 1959, p. 4.

[97] See as examples: (1) DA, "1956 AAP," Annex 1, Inclosure 2, Chart 2, and Annex 2, Appendix A, p. 2. (2) DA, "1957 AAG," p. 4; Annex 1, Inclusion 1, Tables I, V, and VI; Annex 1, Appendix D, pp. 1–3. (3) DA, "1958 AAG," Annex 1, Appendix G, pp. 1–2. (4) Weinert, p. 13. (5) Infantry School, "1956 Instructors' Conference," pp. 104, 125–26.

[98] Major General Hamilton H. Howze, Director of Army Aviation, ODCSOPS, "Notes from the Pentagon: The Use of the Helicopter as a Reconnaissance Vehicle," *USAAD*, July 1957, p. 4.

[99] (1) Collier, p. 26. (2) "Army Operations in Limited War," pp. 44, 45. (3) Wheeler, "Strategic Mobility," pp. 8–9.

[100] Collier, p. 30.

[101] (1) Brigadier General Carl I. Hutton, Commandant, Army Aviation School, "An Air Fighting Army?" *USAAD*, July 1955, pp. 2–3. (2) Brigadier General Carl I. Hutton, Commandant, Army Aviation School, "Future Prospects of Army Aviation," *AID*, November 1956, pp. 18–29. (3) Colonel Jay D. Vanderpool, Director, Combat Developments Office, US Army Aviation School, "Aerial Vehicles in the Ground Role," *MR*, October 1958, pp. 59–65. (4) Howze, "Last Three Years," p. 58.

[102] (1) Claude Witze, "Industry Faces Army Aviation Challenge," *AW*, 26 December 1955, pp. 15–17. (2) Henry T. Simmons, "What the Army Wants: Heavier, Handier Planes—Plus Bigger Air Role," *AA*, 19 December 1955, pp. 25–26. (3) "Front and Center: Exponent of Air Mobility," *Army*, February 1956, p. 15.

[103] (1) Colonel Jay D. Vanderpool, "Attack Helicopter: The Key to Army Air Mobile Operations, A Report for the Blue Ribbon Defense Panel," Fort Rucker, Alabama, 28 February 1970, pp. 11, 16–21. (2) US Army Military History Institute, Senior Officers Oral History Program, Project 83-12, Jay D. Vanderpool, Colonel, USA Retired, Interviewed by John R. McQuestion, Lieutenant Colonel, USA, (Carlisle Barracks, Pennsylvania, 1983), pp. 155–63.

[104] (1) Colonel Jay D. Vanderpool, "We Armed the Helicopter," *USAAD*, June 1971, pp. 2–6, 24–29. (2) Charles O. Griminger, "The Armed Helicopter Story, Part II: Vanderpool's Fool," *USAAD*, August 1971, pp. 14–18. (3) "Office of Combat Development: 'Mission Blue Sky,'" *AA*, 10 March 1958, p. 50. (4) Weinert, pp. 89–94. (5) William Vance, "How the Army got its Shooting Helicopters," *National Guardsman*, May 1963, pp. 2–4, 30. (6) "Army Evaluates Armed Helicopter Units," *AW*, 4 November 1957, pp. 81, 83.

[105] (1) Major Leonard S. Lee, AGC, Assistant Adjutant General, (for the Commander) Headquarters, Continental Army Command, Fort Monroe, Virginia, Memorandum to Commandant, United States Army Aviation School, Fort Rucker, Alabama, "Armed Helicopter Mobile Task Force (U)," 29 November 1957. (2) "Army Tests Heavily Armed Sky Cav 'Copter; 42 Rockets, 9 Machine Guns, 2 Cannons," *ANAFJ*, 2 November 1957, p. 4.

[106] (1) DA, "1957 AAG," Annex 1, Appendix D, pp. 1–3. (2) DA, "1958 AAG," Annex 1, Appendix B, p. 4. (3) First Lieutenant, R. A. Harrington, Jr, Adjutant, CO, 31st Infantry, Memorandum to Director, CDO, "ACR Organization," 24 November 1958. (4) Weinert, pp. 94–96.

[107] Wilber M. Brucker, Secretary of the Army, "Look at the Army Whole," *Army*, December 1957, p. 27.

[108] Brigadier General Carl I. Hutton, "Army Aviation in the Modern Army," *NDTJ*, March–April 1959, p. 65.

[109] (1) Weinert, pp. 130, 139. (2) Charles O. Griminger and Aaron Rose, "Evolution of Armed Helicopter and Future Design Requirements," Fort Rucker, Alabama, 1968, p. 199.

[110] (1) Captain Thomas R. Hill, "ARST," *USAAD*, May 1960, p. 2. (2) Lieutenant Colonel John W. Oswalt, "Low and Slow and with Fire Power," *Army*, November 1960, pp. 56–57. (3) von Kann, "Mobility Now," p. 40.

[111] Army Aircraft Requirements Review Board, Memorandum to the Chief of Staff, United States Army, "Army Aircraft Requirements Review Board," 10 March 1960, Section VII, "Additional Considerations, Inclosure 1, "The Requirement for Air Fighting Units."

[112] Ibid.

[113] (1) Weinert, pp. 30–31. (2) Tolson, *Airmobility*, p. 9.

[114] Seneff, pp. 46–47.

SOURCES FOR TABLES

5-1. (1) DA, "1956 AAP," Annex 1, Inclosure 2, Chart 1. (2) DA, "1957 AAG," Annex 1, Inclosure 1, Table I; Annex 1, Appendix B, p. 6. (3) DA, "1958 AAG," Annex 1, Appendix B, p. 6; Annex 1, Appendix G, p. 1. (4) Department of the Army, Office, Deputy Chief of Staff for Military Operations, "Army Aviation Guidelines FY 1961 – FY 1970," 10 May 1961, Annex C, p. 159.

5-2. (1) Slavich, pp. 88–93. (2) JATB, "1953 Conference," Appendix A14, pp. 177–82; Appendix B10, p. 245; Appendix B11, p. 252. (3) Infantry School, "1957 Instructors' Conference," p. 27. (4) Army, "1957 Airborne Conference," p. 16; Appendix B3, p. 252. (5) Kinzer, pp. 45–48.

6

Structural Handicap of Army Aviation

Because of its organizational structure, Army aviation suffered from a dysfunctional personnel system. The promotion and career structures for aviators were poor at best. Army aviation's bureaucratic structure within the Army staff system was also not ideal. Whether the aviation decision-making process was also dysfunctional is a debatable point. In both cases the lack of an aviation branch played a part. The Army did attempt to create an aviation branch, but it failed.

The Aviation Branch Issue

There was one main argument for creating an aviation branch, which was to lend bureaucratic weight to Army aviation. The main argument against this was the need to have aviators maintain a sense of balance in aviation and ground matters. Specialization in aviation matters would obscure their understanding of ground matters. While this argument prevailed and the Army did not establish an aviation branch, pressure to centralize the structure of Army aviation was ever-present and sometimes seemed irresistible.

Branch Arguments: Generalization Versus Specialization

On its creation in June 1942 Army aviation was an organic part of the Field Artillery. The 1945 extension of Army aviation to other branches adhered to this principle. New users included the Infantry, Armor, Engineers, Signal, Tank Destroyer, Cavalry, and Transportation Corps. (The Army later abolished the Tank Destroyer and Cavalry branches.) There was no separate aviation branch or centralized organization. This obviously had some side effects, especially after the aviation expansion resulting from the Korean war. The lack of a single aviation agency to manage or coordinate burgeoning aviation matters became an anomaly.

When Ridgway instigated a comprehensive review on 4 September 1954 (see Chapter 4), elements within the Army high command made their first attempt to create an aviation branch. Because of its wide-ranging nature, this review involved many Army agencies. The branch issue in particular generated an intense internal controversy. The Army therefore held a conference on this plan on 7 December. Notable participants included General John Dahlquist (Chief of Army Field Forces) and Major General R. M. Montague (Deputy Chief of Combat Developments, OCAFF). Major General Paul Adams (Deputy Assistant Chief of Staff, Operations, G-3) was the representative for the Department of the Army. The main arguments for, and against, an aviation branch emerged during this conference.[1]

Major General Paul Adams was an advocate of an aviation branch. He argued that an aviation branch was necessary since Army aviation was becoming large and complex, and needed its own bureaucratic weight. Other advocates of branch status included Army aviation officers from the Department of the Army staffs who drew up the initial plan.[2] Three main Army elements vigorously opposed this idea. The Field Artillery opposed it because it feared a complete loss of control over artillery observation aircraft similar to that before World War II. The Transportation Corps opposed aviation branch status for a different reason. Colonel Neely presenting a Chief of Transportation paper proposed putting all aviation into the Transportation Corps. However, Brigadier General Carl Hutton and other branches opposed this move. Hutton, himself an artillery officer and representing the Aviation School, also opposed separate branch status. He argued that the Army should accept aviation knowledge and skill as a general qualification, rather than confined to one specialization.[3] Those who shared Hutton's view argued that generalization, otherwise known as "dual proficiency," was necessary for aviators to maintain concern and awareness of ground issues. Also, they argued that ground officers would be more aware of the potential of Army aviation. Another argument against branch status was that it would invite opposition from the Air Force.[4] It would also provide the Air Force with an identifiable organization that it could assimilate into its structure.[5] Thus one can sum up the argument in 1954–1955 by saying that the branch proponents at the Army high command saw the bureaucratic need. Anti-branch advocates inherently took the "generalist" view as a way of making the Army air-minded. They also feared the loss of Army aviation to the Air Force.

Ultimately, the argument for generalization won the day in 1954–1955, and the Army did not create a separate aviation branch. Army aviation therefore remained organic to the seven user branches, with each using arm and technical service retaining jurisdiction over its aircraft and aviation personnel. This not only created career problems for aviators, it also meant that the organs of decision making remained parceled out among the various branches (see below). Not surprisingly, the issue of an aviation branch resurfaced several times during this period.[6]

The Increasing Pressures for Centralization

Pressure to centralize Army aviation remained strong even though branch opponents won the philosophical arguments. While the creation of a branch failed, other schemes to centralize Army aviation were more successful. For example, the Korean war expansion outgrew the Air Training Department of the Artillery School, Fort Sill, Oklahoma. This led to the establishment of a separate Army Aviation School, also at Fort Sill, in March 1953.[7] Later, the Aviation School moved to Fort Rucker and became part of the Aviation Center. This became a central focus of aviation activities.

The Army did not universally welcome this trend. A plan to centralize aviation personnel career management was only partially successful because of this opposition. In 1950 the Department of the Army, G-1 (Personnel) formulated an Army Aviation Officer Career Program. However, in 1951 the Army suspended its implementation pending the end of the Korean war. Ridgway revived this program when he issued the 1955 directive to reorganize Army aviation.[8] However, it became a watered-down version of the original plan because of opposition from anti-branch proponents such as Hutton. The original plan involved removing all aviation personnel files from the various branches and putting them into one aviation file. Hutton called this plan a branch in all but name.[9]

On the issue of aviation tactical organizations, centralization did occur, but over heavy opposition. During Korea, it became Army practice to pool together division and corps aircraft, pilots, maintenance, and supply. Centralization of organic aircraft into a single unit eased maintenance and administrative supervision. This measure became peacetime practice in CONUS and Europe even though it was a highly contentious issue. Hutton, an artillery officer in the Eighth Army during Korea, alleged that certain officers arrived at the decision to pool surreptitiously. There was also other strong opposition to centralizing divisional aviation assets into one unit. The loss of organic control would reduce the close coordination and empathy between aircraft and ground units.[10]

The Pentomic reorganization, beginning in 1955 and 1956, formally established division aviation companies. The Field Artillery opposed this division-wide consolidation because the Division Artillery Commander lost direct control of one of his most important target finding means. In the end the arguments for centralization prevailed, to the chagrin of the Field Artillery. As compensation, the companies contained artillery flights for the direct support of division artillery, with about the same number of aircraft as previously organic to the artillery. Nevertheless, the Field Artillery continued pressing the Army to make the artillery flight organic to division artillery again. The Field Artillery's position was that this return would improve the control, communication, and training between the artillery flight and artillery units, despite the increased maintenance complexities. By 1958 there was considerable support for returning aircraft to the organic control of those units having a continuous requirement for them. By early 1959, ODCSOPS prepared a study on Army aviation that bolstered the position of returning the artillery flight to the organic control of division artillery.[11] The Combat Development Objectives Guide that outlined a policy of integrating Army aircraft at the lowest possible echelon reinforced this reversal of the centralization trend.

This abhorrence of centralization was due to the connotation of a new "Air Corps" that would lose interest in ground support roles and "fly away." For the Field Artillery, one worry in particular was the growing sophistication of pooled aircraft and the diluting of observer skills. One officer pointed out that observation for artillery fire did not need instrument qualifications that were necessary for, say, courier work.[12] Therefore the concept of organic aviation vested in the various branches and technical services had its strong advocates.[13] They saw the centralization of tactical organizations as the first step towards another aviation branch. Moreover, centralization did not mean unity of command, even though this was the intended purpose. There was some confusion over the duties, functions, and responsibilities of the Aviation Company Commander and the Division Aviation Officer.[14]

One aviation officer, Colonel John Tolson, summarized and explained the implications of centralization in 1959. The trend to centralize control of aircraft at higher headquarters was in the interest of economy and possible ease of maintenance, Tolson stated. He said the cost lay in reduced tactical availability to the combat commander and that this imperiled the tactical potential of Army aviation. Commanders with combat responsibilities, he claimed, were losing command control by evolution. He and the Aviation Center believed that centralization at some level would always be desirable. However, he said the Army must resolve whether that command level was at company, battalion, battle group, combat command, division, or army. This was particularly true as the Army as a whole indicated a necessity to increase its aviation effort. Tolson added that, historically, centralization had tended toward specialization. This specialization eventually created special arms or services. If this centralization or pooling continued, he believed the Army would either end up with another corps of aviation, or a technical service would operate the Army's combat aerial vehicles. This would mean a loss of command control by lower combat unit commanders. This, Tolson pointed out, was for the aviation users to decide.[15]

AVIATOR CAREER PROBLEMS AND DISCRIMINATION

The lack of an aviation branch, the policy of generalization, and the immature tactical and bureaucratic organizations presented several serious problems for Army aviators, who were all officers before 1955.[16] The problems included discrimination against aviators by their basic branches, lack of a clear career path, poor career opportunities, and the difficulties and high cost of maintaining dual proficiency. Some of these problems led to further difficulties. The lack of a clear career path and poor career opportunities, for example, led to a shortage of volunteers for flight training. The expansion of Army aviation also led to a shortage of command and staff level officers to fill new positions. The Army did not resolve all of these problems during this period.

Aviator Discrimination

One example epitomized the Army-wide pariah status possessed by Army aviation in its early days. In 1954 a colonel who was the Chief of Policy in ODCSPER managing personnel matters refused to change a policy to allow West Point Military Academy graduates to enter flight training directly. His reason was that a West Point graduate was too valuable to waste on flight training. It took the combined surreptitious bureaucratic moves from Colonel Robert Williams, then in the Aviation Branch under G-3, and the Assistant Chief of Staff, G-3, General James Gavin to have the Chief of Staff (then Ridgway) approve this change. This one act had a significant long-term consequence because it made a significant difference in the caliber of aviation officers.[17]

However, throughout the decade, individual user branches attached pariah status to their aviators. In January 1957 General Howze advised unit commanders not to look upon their aviators as a "separate character out of the mainstream of Army activity." Instead he asked them to see the aviator as an Army officer with a special skill.[18] In late 1958 an Armor officer, Major John Stockton, pointed out three "facts of life" regarding Army aviation. One of them was an attitude from some ground and air officers that aviation was "a separate breed of cat." Stockton speculated that this was because aviators as a whole had less inclination to group discipline and group effort than other soldiers in the Army. He attributed this to the technique of operating an air vehicle, putting a premium on independent thought and action.[19] As a result of this discrimination, many Army aviators were complaining that they were being "kicked around" by their branch units.[20]

Career Structure Problems

The obscure career path or the lack of opportunities associated with Army aviation was in part because aviators competed with nonaviator officers on unequal grounds. While ground officers served entirely in ground duties, Army aviators had to serve in both ground and air positions. At first the policy was for an aviator to serve for a three-year minimum of ground duties after a five-year maximum of flying duties.[21] Later the Army changed this to one-year minimum basic branch ground duties in each rank. This policy ensured that the officer aviators could remain proficient, qualified, and experienced in aviation and in their branch of service, or dual proficiency. However, this theory had several practical problems. Colonel Robert Neely, the Assistant Chief of Transportation (Army Aviation), neatly summed up the first problem in 1954:

> In theory, rotation is fine, but in practice it has its faults. Upon being rotated to ground duty the Aviator is in fact in competition for good efficiency reports with his contemporaries who have spent their full time on "ground duty." On the other hand the increase in the size and complexity of the aircraft and the corresponding increase in time that an aviator must spend on aviation matters, even to meet the minimum safety requirements, has tended further to prevent rotation to ground duty. Yet unless he is rotated to ground duty his chances for promotion are lessened because of his lack of experience in command and other requirements in his basic arm of the service.
>
> The difficulty seems to be that experience and efficiency in aviation matters are not recognized as qualifications for increases in responsibilities and rank, and that the opportunity to command aviation units of appropriate size is generally lacking. . . . For the Army Aviator there is no clear career ladder for advancement.[22]

The one-year period for ground duty also included going to branch and higher command schools. This did not give much time actually serving with field units. Moreover, an Army aviator's flying proficiency suffered while on ground assignment to retain branch proficiency. In most cases, aviators were able to accumulate only the minimum of 80 hours flying time per year.[23] This meant that when they returned to aviation assignments afterwards, there was a temporary loss of flight proficiency. Despite much rhetoric from the "generalist" faction, this was a very difficult problem to solve. General Howze suggested that the unit commander should provide aviators the ground command opportunities. Howze added that it was up to the individual aviator to get promotion beyond the aviation company level. Therefore aviators had to acquire what Howze called a "double competence."[24] Moreover, the practice of dual proficiency was not benefiting the Army as intended. Army aviators frequently received assignment to support branches other than that the aviator came from. Infantry officers who ought to have been flying with infantry units were flying with artillery battalions, and Artillery officers were flying for infantry battle groups. This was taking its toll in efficiency of operation and good personnel management. Situations like these were contributing to a lack of effective support.[25]

The uncertain career path and pariah status led to an aviator shortage in the mid-fifties, when the massive Army aviation expansion began to take off.[26] In September 1955 the Vice Chief of Staff, General W. B. Palmer, published an open letter calling for more applicants for aviation training. The Army needed an application rate of around 300 volunteers per month, for a monthly intake of 200 officers. (Normally one-third of all applicants did not meet the accepted requirements for flight training.) In short, the Army needed an application rate of 3,600 officers per year, or 9 per cent of all eligible lieutenants. The problem was of quality as well as quantity. Palmer said that newly commissioned officers without troop experience made up around 60 percent of the officers entering flight training. Palmer explained: "The fact that relatively few eligible officers volunteer for flight training while on duty with troops is attributable largely to the reluctance on the part of the commanders at all levels to lose officers from their units, and the effect this discouragement has upon the eligible officers of their commands."[27] This reluctance was due to a temporary loss in efficiency that a unit would suffer because of the turnover of officers. Palmer's call, initiated by the Chief of Staff himself, was therefore to encourage more experienced lieutenants serving with troop units to apply. This was a serious problem because the shortage of experienced officer aviators would undermine the long-term well-being of Army aviation and the quality of officers promoted to higher rank.

To improve the prospects of career advancement, the Army high command decided in early 1955 to centralize the career management of Army aviation officers. This was a critical step to encourage more prospective flying volunteers and calm current aviators previously unsettled by the lack of a clear career progression path.[28] The Deputy Chief of Staff for Personnel was responsible for staff supervision of the program. The Career Management Division of the Adjutant General's Office served as a centralized monitor of all aviation officer assignments. The various branches that employed Army aviation still made assignments of aviation officers for their respective branches. However, the centralized monitoring by the Career Management Division ensured the quality and quantity control of branch immaterial aviation assignments, such as high-level staff officers.[29] By 1956, General Howze could claim that Army aviation had "no difficulty whatsoever in getting an adequate quantity of trainees."[30]

Maintaining Dual Proficiency

Another problem area was the prohibitive cost associated with the policy of having dual-proficient officer-aviators. Aviators on rotation to ground duties still drew flight pay and had to maintain minimum flight proficiency. This meant paying a high price to keep expensively trained and sorely needed officers away from aviation duties. Brigadier General von Kann stated that it took one year to establish close cooperation between the ground and aviation units. The turbulence in aviation personnel no doubt put an additional strain on training and readiness. Moreover, the officers had to go through the complete fixed wing training and a period of duty with tactical organizations before becoming eligible for helicopter training.[31] This increased the difficulty of obtaining adequate numbers of helicopter pilots.

The exclusive use of officer-aviators also created certain complications of command and administration. Major Stockton pointed out that the ROCID infantry division as a whole had one officer for every 16 enlisted men. However, in the division aviation company this proportion was only one-to-two. This officer-intensive speciality was demeaning the role of aviation officers and creating problems of career advancement.[32] Therefore the Army had to cut down on the inefficient use of costly trained aviators and improve the officer-men ratio. To do so, the Army initiated a scheme to train enlisted men with civilian pilot licenses directly as cargo helicopter pilots.[33] Congress supported Army plans for this scheme as early as 1953.[34] Later the Army expanded this scheme to include the flying of other aircraft types.[35] Enlisted men would graduate from flight training as warrant officers, whose main criterion was to be proficient in basic and tactical flying. No ground rotation duties were necessary for them. The Army still needed officer aviators for flying aircraft in branch-specific applications requiring branch qualification as well as command and leadership abilities.[36] Nevertheless, the use of warrant officer aviators would ease aviator shortages, the officer-men ratio, and the career path problems. The Army's aim was to have around 42 percent warrant officer aviators in Army aviation.[37]

Senior Aviation Officers

The rapid organizational expansion of Army aviation generated a shortage of higher-ranking aviation officers.[38] Also, because of its previously small size, Army aviation had a "private club" image, with many aviation staff officers knowing one another. The Army solved this by initially bringing in and inducting nonaviation, mainstream general officers and colonels into the top ranks of Army aviation. This scheme had the support of senior officers such as General Howze. The aim of this approach was to give Army aviation a broad outlook and to coopt the Army mainstream into Army aviation (and vice versa). The officers could defend and promote Army aviation and air mobility bureaucratically without incurring accusations of empire building and parochialism. Moreover, they might be better at manipulating the staff system. When aviators with the necessary maturity of outlook become available later on, the Army would promote them to higher rank. Another aim of this cooption program might have been to reduce the pariah status of Army aviation among the user branches. By giving flight training to promising officers, the Army was indicating the coveted nature of possessing aviation knowledge.

The Army carefully selected two classes of middle-ranking officers (lieutenant colonels and colonels) for induction in flight training. One class started on 6 September 1955 and another in 1956. Promising promotion prospects were the main selection criteria for this program.[39] The Army gave flight training to mainstream general officers individually. The success of the middle-ranking officer program is difficult to assess. Of the 24 selected and trained, only a handful became prominent general officers later. Perhaps the general officer induction program was more successful. In 1955, Carl Hutton, Commanding General of the Army Aviation Center, was the only general officer with flight training. (General Howze began fixed wing training when he moved into the aviation staff job in 1955.) By 1960 this total reached 11, with General Paul Adams the ranking officer. He became the commanding general of the XVIII Airborne Corps, then CONARC and then Strike Command (STRICOM).[40] Another notable officer was Major General Herbert Powell, Commandant, Infantry School. In 1959 he became the Deputy Commanding General of CONARC and was the top-ranking aviator in the Army. On 1 October 1960 he became the Commanding General of CONARC.[41]

Other general officers inducted into flight training included the successors of Howze and Hutton. Major General Bogardus Cairns succeeded Hutton in June 1957, but died in an aircraft crash on 8 December 1958. He had a cavalry background, and was previously the Deputy Chief of Staff for Operations, CONARC. He attended fixed wing and helicopter flying courses just before replacing Hutton.[42] Major General Ernest Easterbrook took over on 16 February 1959. (Note how the rank of the Commanding General of the Army Aviation Center had risen from Brigadier General. This may have signaled the growing stature and importance of the post.) Previously Easterbrook was the Director of Army Aviation, with the rank of Brigadier General. He had succeeded General Howze on 16 December 1957. Brigadier General Clifton von Kann replaced Easterbrook as Director of Army Aviation on 20 July 1959 and later rose to the rank of Major General. Before that, he was the Assistant Division Commander of the 82nd Airborne Division, which Major General Howze commanded after leaving his post as Director.[43]

Many senior officers inducted into Army aviation also received a high press profile for doing so. For example, pictures of senior officers receiving flight training, or with flight training already, appeared in the *Army Navy Air Force Journal* in 1957. They included Hutton, Brigadier General Bogardus Cairns, and Colonel John Edmunds, Assistant Commandant, Army Aviation School.[44] Picture profiles of Brigadier General James King, Deputy Commander, Armor Center, and Brigadier General William Bunker, Commanding General, TSMC, also appeared with captions on activities related to Army aviation.[45] Furthermore, the journal published a photograph with Taylor, General Willard Wyman (Commanding General, CONARC), and Hutton together during the Army Aviation Symposium on 5–7 June 1957.[46] If nothing else, this press campaign signaled that flight training may be a path to higher office and promotion. Even civilian officials received flight training. For example, Dr Harold Brown received flight training in the fifties while he was a civilian official in the Department of the Army.[47] He later became Secretary of Defense in 1977.

However, the officer induction program was not a panacea. While this did indeed work for the specific purposes outlined above, it did not solve many other organizational problems. Also, while the evidence is not clear, the induction program might also have affected negatively the promotional prospects of existing aviation officers. Inducted officers would fill the higher offices that otherwise would have gone to career aviators. At a time when a shortage of lucrative jobs for Army aviators was a problem, it would not be surprising if the program caused some problems while solving others.

THE EFFECTS OF DECENTRALIZED DECISION MAKING

The lack of an aviation branch or a central aviation staff agency resulted in a devolved decision-making process for aviation issues. While it is natural to expect this state of affairs to have been detrimental to Army aviation and air mobility developments, the real effect of this system is difficult to assess.

Bureaucratic Structure Within the Army Staff

The Army Policy Council approved the creation of centralized aviation agencies in 1955. This included the Army Aviation Center and the Army Aviation Division under G-3 (see Chapter 4). Despite this, the basic problem of Army aviation decision making remained unresolved. This process remained devolved in nature. Almost without exception every Department of the Army agency had some responsibility in the field of Army aviation (see Table 6-1).[48] Most major organizations outside the Department of the Army also had a special aviation staff section functionally analogous to the Director of Army Aviation. Problems plagued this devolution of bureaucratic control. There was some confusion over the role of different agencies, especially when responsibilities clashed or overlapped.[49] Some commentators saw this aviation structure as a handicap.[50]

The decision-making process involved the coordination of interested agencies. Alain Enthoven, a civilian official in the Kennedy-Johnson administration, said of this decision-making system:

> [I]t is very difficult for an individually promoted idea to reach the Chief of Staff level through normal Service channels. Before this can happen, a proposal must have the agreement of all interested parties. Extensive lateral and vertical coordination is a common feature in both the Service staffs and the Joint Staff, with each different agency exercising its "chop" as the price of its concurrence. As a result, the original idea is likely to be bargained and negotiated until a bland compromise is reached that is acceptable to everyone.[51]

Rather than promoting cooperation, this decision-making system was also engendering rivalries amongst the combat arms and technical services over aviation issues.[52] The structural issue was especially contentious since the combat arms perceived that the Transportation Corps dominated Army aviation bureaucratically.

Table 6-1
Major Division of Responsibilities, Army Aviation (1960)

Agency	Responsibility
Director of Army Aviation, ODCSOPS	Overall staff supervision and coordination of functions relating to Army aviation. Principal focal point for coordination of Army aviation with other Army Staff agencies, other military services and government agencies.
Deputy Chief of Staff for Personnel	Control and manage all Army manpower resources. Army aviators and related aviation-skilled officers and men administered generally through normal personnel procedures. Has a special aviation officer in the classification and distribution division to exercise staff supervision for aviation manpower management.
Deputy Chief of Staff for Logistics	Supervise logistic responsibilities in the areas of procurement, requirements, supply, maintenance, and planning. Include activities of the technical services with direct responsibilities for supporting Army aviation program.
Chief of Research and Development	Planning, coordinating, directing, and supervising all Army research and development. Air Mobility Division carries out responsibilities for Army aviation.
Office, Assistant Chief of Staff for Intelligence	Has an Army aviation staff officer responsible for all intelligence aspects and application of Army aviation. Includes Army requirements for weather service and its coordination with the Air Force.
Office, Chief Signal Officer	Following divisions responsible for avionics and combat surveillance: Combat Developments and Operations; Research and Development; Procurement and Distribution; and Personnel.
Office, Chief of Transportation	Deputy Chief of Transportation for Army Aviation advises on the status of all elements of the Army aviation program that fall within the responsibility of the Chief of Transportation. Obtains from DAA, ODCSOPS, short- and long-range operational requirements for Army aircraft by type. Transportation Corps has major responsibilities for procurement and maintenance of Army aircraft.
CG, USCONARC	Special responsibilities in aviation handled by Deputy Chief of Staff for Materiel Developments and Deputy Chief of Staff for Combat Developments. These offices work directly with US Army Aviation Board and US Army Aviation School in developing matériel and formulating doctrine and concepts.
Army Aviation Center	Army Aviation School has, besides basic training mission, a Department of Combat Developments to project future organizations, doctrines, and concepts. For testing Army aircraft and associated equipment: US Army Aviation Board (service test); the Transportation Aviation Test and Support Activity (TATSA); and the Signal Corps Aviation Test and Support Activity (SCATSA) (concurrent logistical testing).

Transportation Corps Domination

Captain Larry Mickel, for example, felt that there was a logistical bias in air mobility doctrine because of the Transportation Corps.[53] However, this was more a myth than reality. The Transportation Corps did not have the crucial control over doctrinal matters (see Chapter 4). The responsibility for proposing Army aviation doctrine was in the hands of the combat arms. CONARC had overall responsibility, with different branches responsible for different aspects. The Transportation Corps had no say in the *Army Transport Aviation Combat Operations* manual, which was the responsibility of the Infantry School. It was not even the proponent for the field manual on logistical support operations, which was the responsibility of the Command and General Staff College. The Transportation Corps was only responsible for the maintenance and supply manual. Thus allegations of logistical bias due to Transportation Corps influence have no sound basis. The Army made more than adequate progress in advancing the tactical aspects of air mobility.

Captain Mickel also felt that the Transportation Corps had the strongest bureaucratic representation in the Department of the Army staff. The Office of the Chief of Transportation (OCT) represented the Transportation Corps's bureaucratic interests at one level below the Office, Deputy Chief of Staff for Logistics (ODCSLOG). By comparison, there were no such representation for the combat arms. Captain Mickel felt that this gave the Transportation Corps a certain bureaucratic weight and influence on aviation matters at the Army staff level uncompensated by the combat arms.[54] There was little to support this. The creation of the Army Aviation Directorate directly under ODCSOPS headed by a general officer from the combat arms was a deliberate move to give air mobility proponents some bureaucratic counterweight. Moreover, most of the aviation staff officers in the Pentagon were not Transportation Corps officers, nor did the Army concentrate them at OCT or ODCSLOG. Indeed, in 1955 the majority of officers were from the artillery branch, including those within ODCSLOG.[55] This may have been a legacy from World War II.

The existence of the Army Aviation Directorate, however, did not mean a central or even strong decision-making agency. There were fewer aviation staff officers under this directorate than in the rest of the Army staff bureaucracy.[56] It was bureaucratically weak. It had a long list of responsibilities, but it did not have real powers of decision making, especially over personnel, matériel, and doctrine. Its only important function was to prepare the Army aviation program for each fiscal year. In a word, its main role was to "coordinate" between the various staff sections and bureaucracies with shared responsibilities.[57]

The Transportation Corps was responsible for a variety of important administrative functions. For example, it was responsible for supply and maintenance, procurement, research and development, and developing the Army's worldwide aircraft requirements.[58] It therefore had control of most agencies related to Army aviation. All ground crews in maintenance and supply units and those assigned to aviation units belonged to the Transportation Corps as well. One would therefore expect OCT and ODCSLOG to oppose any bureaucratic changes that would make it lose control over aviation units, agencies, and personnel, and the responsibility for their budgeting. It would be a major reduction in budget allocation and hence a considerable loss in influence and prestige. This guarding of its own self-interest would explain much of the Transportation Corps' attitude towards certain air mobility proposals.

For example, despite recognizing the importance of tactical air mobile operations, the Transportation Corps was against the idea of air fighting forces that would make Army aviation take on a combat orientation. The Transportation Corps and ODCSLOG opposed the ARMAIR Brigade concept because they had the most to lose. Not only would control of aviation companies and personnel go to the combat arms, it would also mean transferring responsibility from G-4 to G-3. General Hutton said in 1955:

> Heretofore the Army has tended to consider aircraft as a means of transporting the soldier to battle. There has been comparatively little development in the area of using aircraft as fighting vehicles. . . . The distinction between the two is fundamental. An airplane as a piece of transportation would logically belong to a transportation corps. An airplane as a fighting vehicle would logically belong to the tactical unit of which it is a part. In one case we would have an air transported Army, and in the other case we would have an air fighting Army.[59]

Moreover, the Transportation Corps had a larger percentage of authorized officer aviators than any combat arm. Not only did the Transportation Corps man and operated the high-priority and high-cost transport helicopter companies, it also commanded the largest aviation units in the Army: the transport helicopter battalions. Aviation officers from the other combat arms at best commanded divisional aviation companies, which were no more than administrative organizations for pooled divisional aircraft. They could not get promotion to command transport aviation battalions, which were true commands. In addition, a Transportation Corps assignment meant an opportunity to fly large aircraft.

As a result of these attractions, a survey around 1961 indicated that 18 percent of the 150 aviators questioned at Fort Benning (the

Infantry Center) had transferred, or intended to transfer, from the Infantry to the Transportation Corps. Moreover, these aviators saw in the Transportation Corps a unifying Chief and an opportunity for a full-time aviation job, with a wide variety of aviation assignments and command opportunities. Consequently, the Transportation Corps would *de facto* become the aviation branch unless the Army made sweeping changes in the aviation organization.[60] This "poaching" of aviation officers further fueled the combat arms' suspicion that the Transportation Corps was trying to take over Army aviation. For this reason the combat arms tried unsuccessfully to make branch-immaterial all the transport helicopter companies. This would allow officers from the combat arms to take up assignments to these units without a branch transfer.[61]

Defending its claim on Army aviation, the Transportation Corps juxtaposed its responsibility for transport aviation units with its other transportation responsibilities. During the fifties the Transportation Corps was responsible for coordinating all methods of Army transportation within the theaters. This would include rail, road, air, sea, and inland waterways. It enabled central control of mass movement to obtain maximum and efficient utilization of transport facilities. The Transportation Corps saw Army transport aviation as an extension of its road and rail system in the theater. Transport helicopters could distribute high-priority items moved strategically by air to the theater.[62] It argued that it was most effective to integrate air, sea, and land transport into a single system. This, of course, meant the centralization of control by the Transportation Corps.[63]

In addition, the Transportation Corps was also responsible for the armored personnel carrier (APC). Each division had a pool of such vehicles. A Transportation Special Staff Section within the division coordinated and centrally controlled all modes of divisional transport.[64] This would include any transport helicopters assigned to the division. Perhaps the Transportation Corps was afraid that losing control of the transport aviation companies would diminish the responsibilities of its special staff section. Moreover, it might eventually lose control of the APC companies to the Infantry as well. This was a dangerous threat to its bureaucratic power base. One would therefore expect it to attempt to preserve the logistical orientation of transport helicopters. This included opposing the arming of transport helicopters for tactical missions.[65]

However, because of the devolved decision-making structure, the helicopter armament program crossed several agencies' line of responsibilities. It lay not only with the Chief of Transportation, but also with the Chief of Ordnance and the ODCSOPS. The Ordnance Corps (including its Ordnance Weapons Command, or OWC) and the DCSOPS were both very supportive of helicopter armament developments.[66] The concept of armed helicopters also had the support of most of the combat arms. The Infantry School involved itself in the development of helicopter armaments and was very keen on adopting the armed helicopters by 1959.[67] The Armor School also had a great interest in armed helicopters, especially for sky cavalry missions. The Command and General Staff College illustrated its support for armed helicopters in the MOMAR I study. The Field Artillery did originally express interest in aerial artillery, but this was primarily the work of Colonel Charles Matheny within the combat developments office. After he left in 1957, the Field Artillery interest in this field ceased, and its focus returned to target acquisition and fire adjustment missions. Despite the opposition from the ODCSLOG and Transportation Corps, the process of developing armed helicopters continued unabated. The Army issued its first development contracts in 1958.[68]

The Research, Development, and Procurement Structure

The Transportation Corps was the agency responsible for aviation research, development, and procurement ever since it took them over from the Ordnance Department in 1952. This came under criticism from the Army's civilian advisers.[69] In 1960 a member of the Army Scientific Advisory Panel, Dr Morrough O'Brien, said:

> It is my impression of the organization of aviation in the Army that it is not equal to the job ahead for Army aviation, especially the organization for research and development in aviation. . . . My recommendation is that the aviation equipment be handled in a manner similar to other Army equipment. . . . Tanks and other types of equipment have a branch to take care of their research and development and procurement, and so forth, and it is my thought that aviation requires the same sort of organization to carry through projects promptly.[70]

This control over research, development, and procurement led the combat arms to perceive a Transportation Corps preference for large helicopters to perform logistical missions. These officers saw such large helicopters as completely unsuitable for front-line use because of difficulty of concealment and excess maintenance. They were also too expensive to risk in the front lines.[71]

However, the problem of aircraft unsuitability originated from the Materiel Requirements Review Panel rather than from the Transportation Corps. Moreover, the combat arms' perception did not reflect the priorities set by the Army high command. The Army deliberately adopted a policy of developing and procuring smaller helicopters after canceling the YH-16 program in 1956. Moreover, the Army did procure more helicopters of smaller size following its own official policy. Therefore the claim that the Army procured larger helicopters because of Transportation Corps preference was not true. It was simply that the HU-1 purpose-built for the Army took a long time before reaching front-line use. In the meantime the Army had to buy whatever transport helicopters other services developed. Some officers even blamed the battle group's lack of a single aircraft on this preference for large helicopters.[72] In fact, the regiment (previous equivalent of the battle group) did have two light aircraft before the Pentomic reorganization. The Army pooled them together when it created divisional aviation companies to ease maintenance problems.[73]

It seems, therefore, that while there were some discord because of the structural handicap of Army aviation, it did not have the effect of preventing important developments in air mobility. Nevertheless, by 1960–1961 the persistent structural problems led to a renewed call for an aviation branch, or at least some sort of reorganization.[74] However, Army aviation leaders of this period, specifically Brigadier General Clifton von Kann, still opposed the idea of creating a separate aviation branch.[75]

NOTES

1 Weinert, pp. 7–8.

2 Captain Larry S. Mickel, Member, Air Mobility Group, Airborne–Air Mobility Department, US Army Infantry School, "Centralization is the Answer," *USAAD*, February 1960, p. 22.

3 Hutton, "Without Trumpet and Drums," pp. 129–34.

4 "Jets on the Way as Army Pushes Its Air Strength," *AFT*, 7 May 1955, p. 9.

5 Mickel, "Centralization is the Answer," p. 22.

6 (1) Weinert, pp. 125–26. (2) "Jets on the Way as Army Pushes Its Air Strength," *AFT*, 7 May 1955, p. 9. (3) "No Branch Has 'Divine Right' to

Monopoly on Flying Machine, General Hutton Asserts," *ANAFJ*, 2 June 1956, p. 28.

7 "Army Aviation School Created at Ft. Sill," *ANAFJ*, 14 March 1953, p. 821.

8 Weinert, p. 19.

9 Hutton, "Without Trumpet and Drums," p. 131.

10 (1) Hutton, "Without Trumpet and Drums," p. 128. (2) Captain James C. Smith, 3rd Combat Aviation Company, Fort Benning, Georgia, "Centralized Operations," *USAAD*, March 1955, pp. 19–25, 29.

11 (1) Aviation School, "Conference on DOAA," presentation by Colonel Cantrell. (2) Captain Theodore L. Prevost, "Blueprint for Army Aviation," *National Guardsman*, September 1957, pp. 10, 29. (3) Captain Theodore L. Prevost, "Keep the Artillery's Observation Plane in the Artillery," *Army*, April 1957, pp. 36–37.

12 Prevost, "Blueprint for Army Aviation," p. 29.

13 Major General H. W. Blakeley, "We Must See with Our Own Eyes," *Army*, March 1961, p. 63.

14 (3) Stockton, "Let's Bring Army Aviation on the Team," p. 31.

15 Aviation School, "Conference on DOAA," presentation by Colonel John J. Tolson.

16 (1) "Army Aviation Expands, More Pilots Are Needed," *ANAFJ*, 21 November 1953, p. 340. (2) "Army Sets New Goal for Aviator Program," *ANAFJ*, 9 October 1954, pp. 151, 178.

17 Williams, "Oral History," tape 2, pp. 10–11.

18 Major General Hamilton H. Howze, "Soldiers with a Double Skill," *Army*, January 1957, p. 30.

19 Stockton, "Let's Bring Army Aviation on the Team," pp. 30–1.

20 "B/G von Kann Sees Army Aviation as Key to Survival," p. 20.

21 Marinelli, p. 51.

22 Colonel Robert B. Neely, Assistant Chief of Transportation (Army Aviation), "Army Aviation," *ANAFJ*, 6 November 1954, pp. 273, 302.

23 Mickel, "Centralization is the Answer," p. 20.

24 (1) Howze, "Soldiers with a Double Skill," pp. 30–33. (2) Hutton, "Without Trumpet and Drums," pp. 141–42.

25 Captain Roger M. Pezzelle, Combat Development Office, US Army Infantry School, "Not Branch, but Balance," *USAAD*, February 1960, p. 21.

26 (1) "The Big Hanger: Army . . . " *AFT*, 21 November 1953, p. 4. (2) "Army Aviation Leaders," *ANAFJ*, 11 December 1954, p. 432. (3) "Army Aviation Faces Expansion; Pilot-Plane Boost Nears," *ANAFJ*, 24 September 1955, p. 1.

27 General W. B. Palmer, Vice Chief of Staff, "Army Aviation Training," *AID*, September 1955, p. 32.

28 "Army Sets Better Deal for 3500 Aviators, Career Program Will Aid Promotions," *ANAFJ*, 5 May 1956, p. 4.

29 "Army Announces New Aviators' Career Program," *ANAFR*, 5 May 1956, p. 2.

[30] (1) Jules Witcover, "Where is US Army Aviation Going?" *ANAFR*, 13 October 1956, p. 5. (2) "Army Aviation School Enrollment Climbing," *AT*, 1 September 1956, p. 31.

[31] Marinelli, p. 50.

[32] Stockton, "Let's Bring Army Aviation on the Team," pp. 30–31.

[33] Kilmer, p. 44.

[34] "Enlisted Pilots, Helicopters Loom Large in Army Plans," *AFT*, 13 June 1953, p. 1.

[35] "Army Seeks Helicopter Pilots," *AT*, 24 March 1956, pp. 1, 10.

[36] (1) "Air Force Steps Up Army Pilot Training," *AFT*, 26 November 1955, p. 2. (2) "Army of Tomorrow Gets Wings Today," *AH*, January 1956, p. 15.

[37] Weinert, pp. 21–22.

[38] Weinert, pp. 19–21.

[39] (1) "Special Course in Army Aviation," *ANAFR*, 13 August 1955, p. 12. (2) The Department of the Army published Circular 621-14, dated 26 January 1956 which authorized these classes. See Tolson, "Wings for Infantry," p. 17.

[40] (1) Rigg, p. 25. (2) Eddleman, "Planning for Army Aviation."

[41] See (1) *ANAFJ*, 23 March 1957, p. 3. (2) Weinert, p. 47. (3) "Army Aviation President Wastes No Time; Concern Expressed Over Wing 'Clipping,'" *ANAFJ*, 13 June 1959, p. 4.

[42] "Brig. Gen. Bogardus S. Cairns," *ANAFJ*, 2 March 1957, p. 4.

[43] (1) Weinert, pp. 41, 201. (2) "B/G von Kann Will Head Army Aviation," *ANAFJ*, 9 May 1959, p. 7.

[44] See *ANAFJ*, 2 March 1957, pp. 1, 4.

[45] See *ANAFJ*, 30 March 1957, p. 11.

[46] See *ANAFJ*, 22 June 1957, p. 4.

[47] Williams, "Oral History," tape 2, side 2, p. 11.

[48] (1) "Who's Who in Army Aviation," *AA*, 10 March 1958, pp. 22–25. (2) "Who's Who in Army Aviation," *AA*, 9 March 1959, pp. 26–29.

[49] Mickel, "Centralization is the Answer," pp. 18–20.

[50] Wayne W. Parrish, "Two Handicaps Stifle Army Aviation's Growth," *AA*, 10 March 1958, p. 7.

[51] Enthoven and Smith, p. 100.

[52] (1) "B/G von Kann Sees Army Aviation as Key to Survival," p. 20. (2) Colonel William G. Van Allen, Chief, Operations and Requirements Division, Office of the Director of Army Aviation, ODCSOPS, "The War Between the Branches," *USAAD*, December 1958, p. 2.

[53] Mickel, "Centralization is the Answer," p. 22.

[54] Mickel, "Centralization is the Answer," p. 18.

[55] Robert M. Loebelson, "Army Takes Another Look at Its Air Arm," *AA*, 15 August 1955, p. 22.

[56] Brigadier General Ernest F. Easterbrook, "Notes from the Pentagon: The Increasing Requirement for Army Aviation Staff Officers," *USAAD*, May 1958, pp. 3–5.

[57] (1) Claude Witze, "Army Creates Aviation Division," *AW*, 21 March 1955, p. 20. (2) "The Army's Organization for Aviation," *Army*, November 1960, pp. 38–39.

[58] United States Army Transportation School, "TC Organization and Responsibilities in the Army Aviation Program (Instructor's Guide)," 5 November 1957, pp. 39–40.

[59] Claude Witze, "Industry Faces Army Aviation Challenge," *AW*, 26 December 1955, p. 15.

[60] Mickel, "Centralization is the Answer," p. 20.

[61] Van Allen, p. 2.

[62] Yount, "Tomorrow's Army," pp. 12–15.

[63] Aviation School, "Conference on DOAA," presentation by Captain Robert N. Tedd.

[64] (1) Yount, "Transportation for Peace," p. 58. (2) Major General Paul F. Yount, "Fifteenth Anniversary Message," *NDTJ*, July–August 1957, p. 58. (3) Lieutenant Colonel Karl H. Zornig, "To Pool or Not to Pool: A Pentomic Infantry Division Makes a Test," *Army*, March 1959, pp. 64, 66.

[65] (1) Van Allen, p. 3. (2) Weinert, pp. 96–97, 125–26. (3) Infantry School, "1956 Instructors' Conference," p. 137. (4) Currey, p. 75.

[66] (1) Lieutenant Colonel Charles O. Griminger, "The Armed Helicopter Story Part III: Armed Helicopters Around the World," *USAAD*, September 1971, p. 11. (2) "Army Can't Stay on Ground, AOA Told," *ANAFJ*, 7 December 1957, p. 31. (3) Weinert, pp. 96–97, 99–100.

[67] (1) Major Leo D. Turner, "Organic Battle Group Aviation; We Can't Afford to Wait," *Infantry*, January 1960, pp. 4–9. (2) Infantry School, "DOAA 65–70," pp. 10–12; Annex F, p. 2; Annex I.

[68] Weinert, pp. 87–116.

[69] (1) Claude Witze, "Army Aviation Buying Fight Looms," *AW*, 10 January 1955, p. 14. (2) Garret Underhill, "The Army's Case Against Itself!," *Esquire*, May 1957, pp. 48–50, 99–100.

[70] Dr Morrough P. O'Brien, Dean Emeritus, College of Engineering, University of California; Member, Army Scientific Advisory Panel, "Success Out of Failure," *Army*, September 1960, pp. 69–70.

[71] (1) Pezzelle, p. 23. (2) Mickel, "Centralization is the Answer," p. 20.

[72] Mickel, "Centralization is the Answer," p. 20.

[73] Pezzelle, pp. 21, 23.

[74] (1) Pezzelle, p. 19. (2) Mickel, "Centralization is the Answer," pp. 20, 22.

[75] (1) "B/G von Kann Sees Army Aviation as Key to Survival," p. 20. (2) "Army Aviation Chief Aims at 'Ghosts'; Hits TAC Take-Over, Separate Branch Talk," *ANAFJ*, 31 October 1959, pp. 1–2. (3) von Kann, "The Real Goal of Army Aviation," pp. 35–37. (4) "Army Plans 1970 Aviation Program," *ANAFJ*, 28 November 1959, p. 26. (5) Steve Tillman, Associate Editor, "The Register Interviews: Brig. Gen. Clifton F. von Kann, Army Aviation is Key to Mobility in Battle," *ANAFR*, 12 December 1959, pp. 16–17.

SOURCES FOR TABLES

6-1. "The Army's Organization for Aviation," *Army*, November 1960, pp. 38–39.

III. The Sixties

7

INTERVENTION CONTINGENCIES AND COUNTERINSURGENCY

The beginning of the Kennedy-Johnson presidency in 1961 heralded a new direction in administration policies. The new policies had a dynamic interaction with the international security environment, leading eventually to radical developments in air mobility. One policy change was in internal management. The new defense secretary adopted new management style and techniques at the Department of Defense that closely scrutinized service programs. This critical reexamination of all military programs made it possible for radical air mobility ideas to surface. Another key policy change was in national security policy. The emphasis was on balance in conventional and nuclear forces. The Army reorganization of its divisions away from their predominantly nuclear orientation was an early confirmation of this trend. This policy meant a build-up of conventional ground forces that generated more aviation requirements and made more money available for Army aircraft procurement. The third important policy change was in foreign policy, where there was a determination to halt the spread of insurgencies around the world. This led to increasing overseas intervention, and the adaptation of air mobility to do so. Air mobility at first supplemented Special Forces in this role. Later, it became dominant as the scale of insurgencies increased. The establishment of the Army Tactical Mobility Requirements Board (or Howze Board) and the subsequent formation of an air mobile division reflected these internal and external developments.

THE ROAD REORGANIZATION

Immediately after the Kennedy-Johnson administration came into office the Army began a reorganization that signalled the end of its predominantly nuclear orientation. The Army discarded the Pentomic structure and adopted a new structure with more flexibility to meet limited war contingencies. The new organizational structure began to take shape after the Vice Chief of Staff, General Eddleman, rejected the MOMAR study in December 1960. CONARC produced a study entitled "Reorganization Objectives Army Division (ROAD) 1965" based on studies done at various branch schools and the Command and General Staff College. On 1 March 1961 CONARC submitted it to the Department of the Army. General George Decker, who replaced Lemnitzer on 31 September 1960 as Army Chief of Staff, approved it a month later. Shortly afterward the Secretary of the Army, Elvis Stahr (23 January 1961 to 5 July 1962), recommended to President Kennedy to abandon the Pentomic organization and to adopt the new concept. President Kennedy approved on 25 May 1961, and the conversion of Army divisions to the ROAD structure began in early 1962.[1] Decker's approval of the ROAD structure in April 1961 also immediately set off a series of actions to activate the first air cavalry troop.[2]

The ROAD structure abolished the battle group and reintroduced the battalion. It also introduced the brigade level that replaced the regiment or combat command of old. The three brigades within a ROAD division did not have fixed subordinate commands, but each would control 2 to 5 maneuver battalions drawn from a pool of 10 or 11 battalions. The infantry division had 8 infantry battalions and 2 tank battalions. The mechanized division had 7 mechanized infantry battalions and 3 tank battalions. The armor division had 6 tank battalions and 5 mechanized infantry battalions. The airborne division had 9 airborne infantry battalions and 1 assault gun battalion. In addition, each division had 5 artillery battalions.[3] The reorganization also increased Army aviation significantly within each ROAD division. Initially each new division was to have 50 Light Observation Helicopters, 49 HU-1, and 4 AO-1, or a total of 103. (A mixture of H-13, H-23, and L-19 continued to serve as interim substitutes, since the LOH was not yet in service.) By late 1962 the Army changed this to 48 LOH, 49 UH-1, 4 OV-1, and 2 U-6 (still 103 aircraft in total). This was around double the number in the Pentomic divisions.

In terms of aviation units, the reorganization enlarged the divisional aviation company to an aviation battalion. Within this there was a general support company and an air mobile company. The division also had an air cavalry company, located within the armored cavalry squadron. Helicopter armament kits (machine guns, rockets, and SS-11 missiles) were an official part of these new units. The reorganization also increased the number of aircraft in nondivisional units, for example in independent brigades and in corps.[4] Each corps had an air mobile (light) company, an air mobile (medium) company, an air mobile (fixed wing) company, and an aviation corps company. The field army had no aviation units of its own.[5] These additions increased a nominal field army's number of aircraft to 2,400.[6]

While these new developments in Army aviation were significant, it was still a relatively evolutionary step. The ROAD reorganization did not create any major air mobile force. At this point the combined effects of new administration policies in a changing international environment had not yet produced the need for an air mobile division.

NEW POLICIES AND CHANGING ENVIRONMENT

Balanced Forces and Special Warfare

The Kennedy-Johnson administration's foreign and defense policies tended to be "symmetrical" responses to the Soviet threat compared to the Eisenhower administration's asymmetrical responses. The new administration would contemplate direct ground intervention if there was a threat to vital interests. The aim was to lessen the reliance on nuclear weapons and therefore reduce the likelihood of a nuclear war. The balance of terror in nuclear weapons between the two superpowers was a factor in this consideration. The policy of balanced forces meant an increase in expenditure for conventional forces. The Army received a substantial increase to modernize its conventional weapons and equipment.[7] Army aviation received a commensurate boost because of this emphasis (see Tables 7-1 and 7-2). The Department of Defense also approved an increase of 2 Army divisions, to a total force structure of 16 divisions. Army manpower increased from 861,964 (30 June 1959) to 1,066,404 (30 June 1962). The Berlin crisis of summer 1961 directly caused these increases. Nevertheless, even after the crisis subsided the Department of Defense made the 16-division force permanent.[8] However, personnel strength settled back down to 975,916 on 30 June 1963.

Table 7-1
Army Aviation Budget, FY 1960–1967, Obligations Incurred, Procurement of Equipment and Missiles, Army (PEMA)

Fiscal Year	Aircraft	Aircraft Spares & Repair Parts	Total Direct Obligations	Percentage of Total
1960	$152,000,000	–	$1,388,000,000	10.95
1961	180,000,000	–	1,755,000,000	10.26
1962	256,339,000	–	2,451,051,000	10.46
1963	238,172,000	$50,383,000	2,353,807,000	10.12
1964	421,648,000	65,137,000	2,311,759,000	18.24
1965	335,861,000	69,660,000	2,002,221,000	16.77
1966	938,060,000	250,769,000	5,308,908,000	17.67
1967	883,843,000	223,484,000	4,805,908,000	18.39

Another central tenet of the policy of balanced forces was to increase unconventional forces. It is true that Berlin, along with the two Cuban crises, was high on the administration's foreign policy agenda in 1961–1962. However, other unconventional threats and instabilities around the world also vied for attention.[9] Khrushchev's open support for wars of national liberation began to affect countries in Southeast Asia, Latin America, and Africa.[10] For example, Laos was high on the administration foreign policy agenda because communist guerrillas threatened to overrun the country. (However, the administration decided not to intervene militarily.) To the Kennedy-Johnson administration, rather than Korea-style or European-style conventional warfare, the most likely conflicts were insurgencies. This was evident from the increasing involvement in Vietnam, even though the insurgency problem there was not very important initially. The development of counterinsurgency capabilities was therefore high on the administration's priorities.

While President Kennedy initially took a personal interest in air

Table 7-2
Army Aircraft Inventory, FY 1960–1968

Fiscal Year	Fixed Wing	Rotary Wing	Total	% Increase	No. Increase
1960	2,860	2,633	5,493	–	–
1961	2,843	2,721	5,564	1.3	71
1962	N/A	N/A	5,648	1.5	84
1963	N/A	N/A	6,001	6.2	353
1964	2,889	3,449	6,338	5.6	337
1965	2,693	4,240	6,957	9.8	619
1966	2,604	5,494	8,098	16.4	1,141
1967	N/A	N/A	9,490	17.2	1,392
1968	2,380	8,085	10,465	10.3	975

mobility, this does not seem to have had a direct impact on developments.[11] The only military issue Kennedy remained directly involved in was the Special Forces.[12] The Special Forces were originally to act as guerrillas behind enemy lines. The administration converted the Special Forces to counter insurgencies instead, leading to their massive expansion. The Special Forces had an interest in Army aviation for counterinsurgency.[13] It usually had an aviation element as part of its deployment package (see below).

The administration sent Mobile Special Forces training teams to Southeast Asia, Latin America, and the Middle East.[14] Such military assistance involving the Military Assistance Advisory Group (MAAG) or Mission came in two phases. In the first phase, the MAAG provided training advice and assistance only. In the second phase it provided operational advice and assistance, where military advisers actively participated with indigenous forces in counterinsurgency operations. In early 1962 the Army was providing Special Forces training in Laos (first phase), while in Vietnam aviation units were participating in counterinsurgency operations (second phase).[15] This counterinsurgency requirement increased as the situation in Southeast Asia deteriorated. The number of advisors and Army aviation units there grew relentlessly after their first introduction in late 1961.

The Kennedy-Johnson administration saw helicopter mobility as an essential part of propping up South Vietnam.[16] The manpower requirement for a counterinsurgency campaign was considerable. A rule of thumb was that government forces needed a ten-to-one superiority. The South Vietnamese regime did not have nearly that degree of superiority. It probably had no more than a five-to-one advantage over the Viet Cong. The hope was that American support in the form of tactical advice, modern firepower, rapid communications, air support, and the growing mobility from armored personnel carriers and air transport, would provide the needed margin of superiority.[17] No doubt this was the reason the South Vietnamese government wanted the United States to help its Army (ARVN) to establish an army aviation force.[18]

It was through the growing importance of propping up the South Vietnamese government that helicopter transports received extra funding and program support from the highest political level. In November 1961 the first contingent of six Army aviation units began deploying to Vietnam. This included three light helicopter transport companies, two of which arrived on 11 December with 33 H-21C and 400 men.[19] One light fixed wing aviation company and a battalion headquarters deployed in January 1962. A medical air ambulance company deployed in March 1962. As the situation in Vietnam gradually deteriorated, the Kennedy administration sent more military advisers, helicopters, and tactical air support. Army aviation units in Vietnam increased as the total number of US military personnel there increased each year.[20] After Johnson assumed power in 1963, the likelihood of direct intervention in Vietnam also increased.[21] By 1965 the Army had deployed 13 company-size aviation units to Vietnam, some of which it stripped from divisions in the United States and other overseas commands. Since they were for helping the indigenous Vietnamese forces, even if some Army divisions deployed to Vietnam the aviation units would not be available to them. These aviation deployments decreased the rotation base required by peacetime personnel policies and increased the repetitive short tours for career aviation personnel.[22] These pressures increased the need to create more aviation units.

The increase in overall Army force structure, divisional aircraft, military assistance programs and overseas deployment all increased the requirement for Army aviation.[23] This led to a shortage of about 1,000 aircraft, 1,200 aviators, and maintenance personnel. The obsolescence of the Army's current aircraft fleet aggravated the procurement situation even further. If the Army procured more existing types of aircraft to fill the requirements, it would end up with more obsolete aircraft to replace. This was a situation the Army was not very happy with.[24] Moreover, this was the basis for the Secretary of Defense's subsequent critical review of Army aviation.

New Defense Management Style

The new Secretary of Defense, Robert McNamara (21 January 1961 to 29 February 1968) and his civilian political appointees introduced the management technique known as systems analysis to the Department of Defense. It was completely different from before because the new style was much more inquisitive and intrusive than merely setting the budget ceiling. The method of system analysis involved a formal procedure called the Planning-Programming-Budgeting System (PPBS). This involved a procedure of open and explicit analysis, review and comment, by all interested parties. The intention was to ensure that the Secretary of Defense would hear all sides of the arguments, with no one bureaucracy holding a monopoly on the information going to him. Its aim was to protect against persistent error or the risk that any group might rig the analysis results.[25]

The management method of systems analysis scrutinized the efficiency and cost-effectiveness of military expenditure. The aim was to require the military services to provide proper justification for large and expensive programs. This close civilian audit of military programs went hand-in-hand with the marked increase in defense spending. Civilian systems analysts played a key role in analyzing the justifications or performing their own independent analysis.[26] A corollary of this management style was to drive for improved efficiency and more value for money, cutting all unnecessary spending.[27] Therefore, along with the massive increase in defense spending, McNamara also instigated a cost reduction program.[28] He said: "Adding a weapon to our inventory is not necessarily synonymous with adding to our national security. . . . This process of choice must begin with solid indications that a proposed system would really add something to our national security."[29]

As a result of PPBS and the cost reduction program, Army aviation came under the personal scrutiny of McNamara and his civilian advisers.[30] McNamara and his civilian staff at first did not push for radical air mobility. They reviewed Army aviation from late 1961 because it was costly and not properly justified. While Army aviation was under scrutiny, civilian officials (colluding with Army air mobility proponents) exploited the opportunity for turning it into a vehicle for radical change.[31] The Marine Corps development in air mobility, being further ahead than the Army's, indicated the possibilities.[32] As a result, in April 1962 McNamara sent two memoranda to the Secretary of the Army asking him to reexamine air mobility radically. He asked for the creation of the Tactical Mobility Requirements Board to do so. He also specified that General Howze was to chair the board because he was a known visionary.[33] McNamara and his staff analyzed the concepts recommended by the Howze Board and asked the Army to form a provisional air mobile division for thorough testing, experimentation, and evaluation.

While the role played by McNamara and his officials was important, most histories of air mobility overemphasize it. The image portrayed is that of "progressive" and "radical" civilians pushing the "conservative" and "reluctant" Army.[34] Alain Enthoven, a prominent civilian official in the Department of Defense, claimed that the air mobility revolution in the sixties occurred because of civilian intervention. He argued that civilian officials brought the air mobility concept to the Secretary of Defense's attention, bypassing conservative service elements. Enthoven therefore concluded that civilian officials' support and active interest played a major role in speeding up the air mobile division's introduction into the Army force structure.[35]

The civilian analysts' willingness to consider unorthodox proposals also helped air mobility development in other ways. They played an important role in putting the issue of vulnerability into perspective, and therefore reduced its controversy.[36] In addition, McNamara approved a temporary increase in personnel to accommodate the new provisional 11th Air Assault Division and the 10th Air Transport Brigade. He also accelerated the procurement of aircraft for the same purpose.[37] Moreover, the new management style played a crucial role in silencing interservice bickering that increased in direct proportion with the radical air mobility proposals. It also allowed the Army more flexibility in procuring aircraft. McNamara, while he did not rescind Wilson's memorandum, was completely disregarding it for policy or planning purposes.[38]

It was true that the new defense management techniques were responsible for *accelerating* air mobility development. However, the Army's role was still crucial. For one thing, the Howze Board had very little direct involvement from the civilian analysts. The board consisted of Army officers and Army civilian advisers. There was also a limit to how radical the OSD officials were. The Howze Board made the radical multi-division proposals (see below). McNamara and his staff doubted whether the Army needed more than one or two air mobile divisions. Even during the divisional tests they were not sure of the exact extent to which air mobility should replace ground mobility.[39]

Moreover, implementing any radical proposals depended entirely on the Army. If the Army had not had sufficient aviation units already in place by 1962, it is doubtful it could have carried out a divisional air mobility test. The resources required for the divisional test were additional to the growing aviation forces deployed to Vietnam, even though the Army was assimilating an increasingly extensive air mobile force structure. It was the Army that established the 11th Air Assault Division and the 10th Air Transport Brigade in 1963. It set up and conducted the divisional test until 1965 under the code name Test and Evaluation of Air Mobility (TEAM). Army officers wrote the test report and recommendations.

Furthermore, the Army high command encountered difficulties in air mobility indoctrination and training during the fifties. Civilians could hardly do better. Before the provisional air mobile division would carry out the tests, it had to receive training to function as a division. Training of aviation units other than the air mobile division was also important. It allowed the widespread use of air mobility in

Vietnam and was instrumental in reversing the insurgency situation, from stage-two guerrilla warfare back to stage-one.

The Army as a whole did not drag its feet and accepted the conversion of one of its division to an air mobile division in 1965. It was the Army that assembled, equipped, prepared, and trained a completely novel division into a combat-ready force. Without the basic building blocks already in place, it is doubtful if the Army could have done so and deployed it to Vietnam eight weeks later. This was especially true since the division had to replace half its troops before leaving for Vietnam because they were ineligible for overseas deployment.[40] Procurement of the necessary aircraft and equipment, devising the multifarious doctrine, and effective training of officers and men for the air mobile division and other aviation units cannot happen overnight. It is fortunate that the Army started to do the groundwork over a decade ago. Civilian officials only accelerated and aided the process of testing and forming the air mobile division. The Army high command's role was still critical in accepting air mobility and implementing it.

THE ARMY AND THE HOWZE BOARD

The Army Tactical Mobility Requirements Board

The Army Tactical Mobility Requirements Board chaired by General Hamilton Howze (or Howze Board) began its work in early May 1962. On 1 September 1962 CONARC submitted its final report (dated 20 August 1962) to the Secretary of Defense. The report recommended 5 air assault divisions, 3 air cavalry combat brigades, and 5 air transport brigades within the 16-division Army force structure. Each air assault division was similar in structure to other ROAD divisions, with 15,029 men. It was to have 3 brigade headquarters, 8 air mobile infantry battalions, 5 artillery battalions, and 1 air cavalry squadron. The aim was to reduce the division's ground vehicles from 3,452 to 1,113, replacing them with 459 aircraft. This included 154 UH-1 tactical transport helicopters (redesignated from HU-1) and 48 CH-47 transport helicopters (redesignated from HC-1). These transport helicopters could move one-third of the division's combat elements in one lift. The reduction of ground vehicles also meant that the amount of strategic lift required was less than for other types of ROAD divisions. The board saw its most applicable areas of employment as, in order of suitability, in an unsophisticated formal war, counterinsurgency, and sophisticated

Table 7-3
Cost Comparison of ROAD and Air Assault Divisions (Million of Dollars)

Division Types	Initial Equipment Cost (CONUS)	5-Year Operating Cost (CONUS)	Total 5-Year Cost
Air Assault Division	282	705	987
ROAD Mechanized Division	155	646	801
ROAD Infantry Division	111	582	693
ROAD Armored Division	182	681	863
ROAD Airborne Division	76	579	655

formal war.[41] The cost of equipping and running an air mobile division was comparatively higher than for other types of divisions (see Table 7-3). However, the intention was to have increased effectiveness to compensate for this. (The official air mobile division formed in 1965 was only slightly different from this proposal.)

The air cavalry combat brigade was a completely air-fighting unit with 316 helicopters, of which 144 were attack helicopters. Its main composition was three air cavalry combat squadrons. The board saw its most applicable areas of employment as, in order of suitability, in a sophisticated formal war, an unsophisticated formal war, and counterinsurgency. The air transport brigade was to have 134 aircraft. Its main role was to reinforce the transport capabilities of the air assault division, or other corps and division units. The corps aviation brigade with 207 aircraft was to perform mainly transport and surveillance functions. The Howze Board also recommended augmenting other ROAD divisions and armored cavalry regiments with additional aircraft. Moreover, it recommended making corps artillery air mobile, and creating a special warfare aviation brigade.[42] The Commanding General of CONARC, General Herbert Powell, to which the Howze Board reported, was very supportive of the report.[43] He was previously the Commanding General of the Infantry Center and had always been a keen supporter of air mobility.[44]

Responses to the Howze Board Proposals

One commentator, Perry Poe, suggested that the tone of McNamara's two memoranda on 19 April 1962 and their tendency to "specify details rather than policies infuriated some members of the Army staff." However, this did not mean that the Army high command resisted the board or its recommendations. Perry Poe said that the Howze Board received the wholehearted support of the

Army.[45] For one thing, General Decker, the Vice Chief of Staff from 1 August 1959 under Lemnitzer, was a supporter of an increased Army aviation program. His view on major military problems was almost identical with General Lemnitzer's.[46] Decker, like his predecessors, placed equal importance on improving air mobility and ground mobility. One of the first speeches Decker gave after becoming Chief of Staff was on the subject of Army aviation, at the National Defense Transportation Association.[47] He told the House Subcommittee on Defense Appropriations:

> The increased range and lethality of the weapons available to us and to our potential enemies require that we adopt wide frontages and dispersed formations on the modern battlefield. These factors, in turn, dictate a need for greatly-improved mobility and for improved combat intelligence—to maintain surveillance of the battle area and to acquire targets presented by the enemy. Our greatest advances in tactical mobility and combat intelligence will be made possible by new forms of air vehicles specifically designed to live with Army forces on the battlefield. Our Army objective is to provide aircraft which can operate from unimproved terrain under all weather conditions, with minimum maintenance requirements—to give us much better capabilities for target acquisition, surveillance, and mobile combat operations.[48]

Aviation remained the number two priority in the Army's budget request to Congress, second only to missiles.[49] In early 1961 the Army was spending 28 percent of its procurement budget on missiles,[50] compared with 11 percent on Army aviation.[51] Moreover, in 1961 the Army requested, and obtained, a significant annual budget boost for Army aviation from Congress.[52] It also received a boost in the number of Army aviators allowed, at the expense of the Navy and Air Force.[53] These increases were in line with the Rogers Board recommendations.

In an article published in October 1962, Decker stated that the Army could afford or profitably use "a few" air assault divisions, especially in Southeast Asia. He was also receptive to the idea of using air lines of communication, both within the theater of operations and from the United States to overseas commands. Moreover, he said that advances towards revolutionary goals normally occurred by short evolutionary steps taken in the hard light of fiscal reality. He said the Army must orient these evolutionary changes to clearly defined and sufficiently demanding longer-range goals. He also suggested taking a hard look at roles and missions for revision to integrate the Army's efforts better with other services.[54]

General Earle Wheeler, who took over from Decker on 1 October 1962, was also supportive of air mobility. Although Wheeler was pro-armor since he had an armor background, Perry Poe said that Wheeler was being "sympathetic and open-minded" to the Howze Board recommendations.[55] The only caveat Wheeler had was that air mobility must satisfy the Army's objective of prompt and sustained combat on land. This was especially important since there were many competing demands on Army resources.[56] In October 1963 General Wheeler stated: "The Secretary of the Army and I are convinced that our recommendations in the airmobility field can significantly increase the Army's capabilities. At the same time, it is prudent to test the concepts thoroughly before proposing their full-scale implementation."[57] Nevertheless, General Wheeler was supportive of the Howze Board and air mobility in his outlook:

> In my opinion the Howze Board performed a unique and distinctive service to the Army and to the Defense tri-service team by exploring the possibilities of applying Army aircraft to the natural evolutionary development of improved firepower and mobility systems for the Army. Our joint tri-service test program will determine the extent to which these concepts will improve Army combat effectiveness. I believe that the time is at hand when we should accept the Army aircraft as a primary vehicle to function with the jeep, the truck, and the tank in ground combat. Where in the past we have used wheeled and tracked vehicles to furnish mobility, protection, and fire support on the ground, now we can look to the Army aircraft for the same function in that dimension of ground combat opening up to us just a few hundred feet over the head of the soldier.[58]

In March 1964, while the divisional tests were still going on, General Wheeler said:

> It is our conviction that the next significant addition to mobility is through Army aircraft. We believe that there are appropriate operations for aircraft of this

> type in land combat situations ranging from counterinsurgency to general war. The air-mobility role must be soundly conceived within the capabilities of men and machines for each type of operation. To misjudge it can be as critical an error as was the Allies' misjudgment of the use of armor early in World War II. The officers and men taking part in the tests now underway at Fort Benning have a mission as important as any in the Army today.[59]

The Secretary of the Army, Cyrus Vance (5 July 1962 to 20 January 1964), was also a staunch supporter.[60] In early 1963 he informed the Secretary of Defense that the Army concurs in the basic proposals advanced by the Board and recommended the adoption of the concept.[61]

THE VIETNAM INFLUENCE

Although McNamara asked the Army to reexamine air mobility radically, this did not mean that the Howze Board's recommendations entirely convinced McNamara and his civilian staff. Later, OSD began to have fewer reservations. The growing Vietnam involvement meant a need to increase mobility to combat insurgents. The exigencies of Vietnam also meant that a larger-scale counterinsurgency force was necessary. The Army high command, with counterinsurgency as a new rationale, turned to air mobility. Thus battlefield needs ultimately affected the direction of air mobility development.

Army Aviation and Counterinsurgency

There was no doubt that the rapidly changing international situation was dictating a change in the Army's outlook and priorities.[62] General Trudeau, the Army Chief of Research and Development, turned his focus to guerrilla warfare as early as 1960.[63] Decker involved himself actively in putting the Army and Army aviation on a better counterinsurgency footing. He said in early 1961: "At present, the level of most active conflict is in the 'low end' of the spectrum—as in Laos, Vietnam, Cuba and the Congo—with emphasis on subversion and insurgency."[64] He noted that missiles and other sophisticated matériel were not the answer. He said the Army was undertaking with great urgency to increase its own and its allies' capabilities in special warfare and counterinsurgency techniques.[65]

By early 1962 the Army was forming four small military task forces organized and trained for what Decker called "cold war areas." These task forces contained engineer, medical, supply, transportation, civil affairs, and light aviation components for accomplishing many essentially nonmilitary tasks. One such cold war task force was at Okinawa, with an orientation towards Southeast Asia and the Western Pacific. One was in Europe, with an orientation towards North Africa and the Middle East. Two were in the United States but oriented on Latin America and Sub-Saharan Africa.[66] Resources and attention devoted to "cold war measures" had therefore increased substantially.[67] Special Forces and their aviation needs were also increasing. For example, on 21 March 1962 the Army activated an aviation unit specifically designed to support counterinsurgency operations: the 22nd Special Warfare Aviation Detachment, Light Aviation Special Support Operations (LASSO).[68] However, the requirements of the Special Forces and their influence on air mobility developments were relatively minor.

It was the operational demand from the large-scale counterinsurgency effort in Vietnam that made the Army convert air mobility for counterinsurgency.[69] This complete reorientation from Special Forces to air mobility was quite apparent in the Secretary of the Army's annual report between 1962 and 1964. Over these three years, air mobility gradually became more important than special warfare as an entry in the report.[70]

The need for more counterinsurgency forces led the Army to consider making conventional forces air mobile enough for this type of operations. Starting from 7 December 1961 CONARC began to consider what augmentation a ROAD division's brigade required to conduct such operations. CONARC addressed three levels: minimum brigade air mobility, complete brigade air mobility, and complete division air mobility.[71] The Army also considered creating dedicated air mobile forces for counterinsurgency. In July 1961 the Plans Division, under the Deputy Chief of Staff for Combat Developments, CONARC, prepared a concept of operations for air mobile forces in an underdeveloped area. On 18 January 1962 the same Deputy Chief directed its War Games Division to conduct a war game of an air mobile division in an underdeveloped area. Major General Thomas Van Natta chose Laos and the surrounding region as the locale. The war games had the People's Army of North Vietnam (PAVN) as the enemy.[72] In September 1961 Van Natta also directed Combat Operations Research Group (CORG) to study the use of helicopter tactical forces. This dealt largely with the armaments appropriate for such a force, the survivability of aircraft used, the required logistics, and the costs involved. The framework for this study was predominantly counterinsurgency, although conventional considerations were also present.[73] All these occurred before the Howze Board convened and around the time when the first Army aviation units went to Vietnam.

Most important of all, on 18 December 1961 CONARC convened a Special Warfare Board chaired by Lieutenant General Hamilton Howze. He was then the Commanding General of XVIII Airborne Corps and STRAC. This board on unconventional warfare, which published its report on 28 January 1962, occurred well before the Howze Board on air mobility. The report had a thinly veiled call for converting three divisions into air mobile divisions for the counterinsurgency role.[74] This meant that Howze not only understood the political and strategic necessity to consider applying air mobility to counterinsurgency. He also understood that the magnitude of the problem required drastic action, such as forming multiple air mobile divisions.

It is therefore not surprising that the Howze Board on air mobility was very aware of this and asked for five divisions. It also delineated three different contingencies: sophisticated formal war, unsophisticated formal war, and counterinsurgency. The first implied a European or Korean-style war, while the second implied contingencies in the Middle East, Latin America, and Africa. The board recommended different types of air mobile forces for each. Board members also visited Southeast Asia, and included a report of this. The board also devoted a section to special warfare and recommended a special warfare aviation brigade with five squadrons.[75] The changing strategic and military developments were therefore having an influential and wide-ranging impact on air mobility development.

By 1963 counterinsurgency considerations had become all-pervasive in air mobility. This change-over from a conventional warfare or nuclear orientation was quite apparent. For example, in November 1961 Brigadier General von Kann discussed the case for air cavalry on the dispersed nuclear battlefield, although he did discuss infiltration and guerrilla units under this framework.[76] Secretary of the Army Elvis Stahr said in 1961 that air mobility "will be a great advantage in any conflict, but indispensable in the event of an atomic war."[77] However, by February 1963 General Wheeler told Congress: "Initial field tests and war gaming of air assault units engaged in performing basic combat tasks indicate that air mobility yields sizeable advantages [in] all types of war. The greatest benefits accrue in the counterinsurgency and limited war situations."[78] In 1964 he added that, in the area of tactical mobility,

> we seek what has been described as the "mobility differential" or advantage over the enemy in means of maneuver. In Viet Nam, for example, the terrain and the environment hold the rate of ground movement substantially to that of the soldier on foot. Thus the addition of the helicopter gives the Army of the Republic of Viet Nam a mobility differential and the advantages that goes with it.[79]

During the 1965 Infantry Instructors' Conference, Major General Robert York, Commandant of the Infantry School, told his audience: "I think counterinsurgency is the biggest and most important problem we face."[80] Major General Edward Rowny said in an article:

> These key capabilities of the air mobile division give it a military effectiveness permitting it to execute many of our current contingency plans. Varied offensive operations could thus be carried out in both stability operations and in conventional war. . . .
> In considering the operational environment, indications are that an air mobile force probably has greater potential than any other type ground force in stability operations, and excells [*sic*] in mid-or low-level conventional war. The capabilities of air mobility in a nuclear war requires [*sic*] further investigation.[81]

By this time, air mobile forces had gained wide recognition within and outside government as more suitable for guerrilla warfare rather than European-style conventional warfare.[82] This came about because of the growing Vietnam war and air mobility operations within it.

Growing Combat Experience and Requirements

Army aviation's involvement in the Vietnam war came in two phases. In the initial learning phase, Army aviators taught Vietnamese commanders and soldiers how to employ helicopter tactics effectively. At the same time Army aviators were learning by experience, trial and error. Even during this early stage and despite being classified, Army aviation operations in Vietnam were receiving official attention at the highest levels.[83] The Army, as early as October 1962, circulated and disseminated its first document on the operational experiences gained by Army aviation units while supporting counterinsurgency operations in Southeast Asia, especially in South Vietnam. It distributed this report widely for aviation unit commanders and their staff officers to analyze, and to assist in doctrinal development and unit training. While the document stated that doctrines previously developed in the fifties were fundamentally sound, some modifications were nevertheless necessary.[84] Apart from official reports, the Army also disseminated the Vietnam and other counterinsurgency experiences in Army magazines such as the *US Army Aviation Digest*.[85]

The second phase had a more extended scope. It involved battalion-size air assaults of selected Vietnamese units. It was during this phase, when the first air mobile division was under test, that the Army made great improvements on the tactical employment of helicopters. The Army studied the variety of tactics used by the Vietnamese and their American support in these air mobile operations. The Army tried to apply everything it learned to the organization and training of its air mobile units.[86]

The testing of the 11th Air Assault Division during Project TEAM, although including a semisophisticated enemy,[87] had conventional operations as its main focus. This was because conventional warfare was the most challenging. The reasoning was that if the air assault division was able to face this challenge, then it could face any lesser problems. Moreover, the 11th Air Assault Division did not need to test properly the application of air mobility for counterinsurgency. The Army and the Department of Defense were already getting direct experiences fed back from Army aviation units deployed to Vietnam. Even the Project TEAM final report admitted that "lessons learned in combat are nearly always the best bases for improved doctrine and equipment. In this instance, adequate proof is at hand in the current successes of heliborne operations in the Republic of Vietnam."[88]

In addition, while the 11th Air Assault Division test was ongoing in the United States, the Army also had an agency out in Vietnam testing air mobility concepts in that environment. The Army Concept Team In Vietnam (ACTIV), headed by Brigadier General Edward Rowny, tested all kinds of equipment and employment techniques for counterinsurgency.[89] Indeed, ACTIV was responsible for 10 of the 41 studies and tests conducted for Project TEAM.[90]

In February 1964 Gene (Eugene) Vidal, Chairman of the Mobility Panel, Army Scientific Advisory Panel, received a briefing at Fort

Eustis, Virginia, from former aviation commanders in Vietnam. In May 1964, Lieutenant General Dwight Beach, the Commanding General, Combat Developments Command, reassembled at Fort Rucker a larger group of officers for a conference. This included five majors and a captain, of whom four were former battalion commanders in Vietnam. Professor Rene Miller of MIT, chairman of the Air Mobility Sub-panel, was present along with Vidal. After the conference Vidal reported to Assistant Secretary of the Army Willis Hawkins.[91] While these discussions centered on operational and technical matters, they nevertheless indicated the importance the Army high command attached to battlefield experience.

The armed helicopter was another example of a war-driven requirement. In the exigencies of war the Army found that helicopter assaults had better protection if escorted by armed helicopters. It therefore formed a new unit, the Utility Tactical Transport Helicopter Company, equipped with 15 new turbine-powered UH-1, for this role of supplying helicopter suppressive fire. The company deployed in September 1962.[92] Armed helicopters became increasingly important because of the Vietnam war.[93] As a result of lessons from Vietnam, the Army started a program for an Advanced Aerial Fire Support System (AAFSS) in 1964.[94] However, the urgency of the Vietnam requirement for a proper armed helicopter meant that the Army settled for an interim aircraft, the AH-1. This helicopter used the same engine and rotor dynamics as the UH-1. The major change was the streamlining of the fuselage design to get more forward speed.[95]

At the end of the TEAM test McNamara and his civilian analysts did not challenge the lack of proper testing in anti-guerrilla warfare. This was unlike the extensive critique they gave the Howze Board report. The experience from Vietnam was providing sufficient evidence of air mobility's application in counterinsurgency operations. Moreover, events were making the formation of an air mobile division inevitable. In early April 1965 President Johnson decided to send combat troops to Vietnam.[96] On 11 June 1965 the JCS recommended sending an air mobile division there.[97] McNamara announced the formation of an air mobile division on 16 June 1965, although he refused to acknowledge that it was for Vietnam. Nevertheless McNamara and the statement released by the United States Information Service on 17 June 1965 suggested close linkage between the two: "Mr McNamara said the airmobile division would make its greatest contribution to combat effectiveness 'when terrain obstacles and undeveloped communications networks could give enemy guerrilla forces an advantage.'"[98]

News reports following this announcement had no doubt that the new capability was for the growing needs of Southeast Asia.[99] Vindication of their guess occurred less than two months later, in August.[100] The 1st Cavalry Division (Air Mobile) was the first divisional unit to arrive in Vietnam. From October 1965 it conducted a textbook campaign at Ia Drang Valley against a North Vietnamese division.[101] The creation of this division represented the final step or ultimate conclusion of the air mobility innovation. It had finally become a normal part of land combat operations.

NOTES

1 (1) Weinert, p. 72. (2) Doughty, p. 21. (3) Lieutenant Colonel Morris G. Rawlings, Director, Combat Developments Office, US Army Aviation Center, "Army Aviation and the Reorganized Army Division," *USAAD*, February 1962, p. 1.

2 (1) Weinert, pp. 139–40. (2) "2nd Air Cavalry Troop (Provisional)," *USAAD*, January 1962, pp. 13–14.

3 (1) Doughty, p. 21. (2) Rawlings, "Army Aviation and the Reorganized Army Division," p. 2.

4 (1) Weinert, pp. 73–7. (2) Rawlings, "Army Aviation and the Reorganized Army Division," pp. 2–4. (3) United States Army Infantry School, "Infantry Reference Data, ROAD," Fort Benning, Georgia, September 1962, pp. 5–6, 46, 48, 50, 120, 124, 126, 154, 174. (4) United States Army Infantry School, "Infantry Reference Data, ROAD," Fort Benning, Georgia, February 1964, pp. 8, 42, 44, 46. This source indicated that each division had 4 OV-1, 47 LOH, 19 utility helicopters, and 31 tactical transport utility helicopters. The airborne division had 6 OV-1, 47 LOH, 25 utility helicopters, and 25 tactical transport utility helicopters. The utility helicopter was probably the UH-1A or UH-1B, while the tactical transport utility helicopter was probably the UH-1D.

5 Department of the Army, Office, Deputy Chief of Staff for Military Operations, "Theater Type Corps Force (Objective) FY 1965 (U)," Washington 25, DC, 2 May 1962, p. 14.

6 Brigadier General Delk M. Oden, Director of Army Aviation, ODCSOPS, Address to the Aviation Writers Association, Palo Alto, California, 24 May 1962, pp. 4–6.

7 (1) Enthoven and Smith, pp. 167–68. (2) Kaufmann, p. 287.

8 Weinert, pp. 33–38.

9 (1) General George H. Decker, Chief of Staff, Address at the National Strategy Seminar, Army War College, Carlisle Barracks, Pennsylvania, "The Military Aspects of the Cold War," 8 June 1961. (2) General George H. Decker, Chief of Staff, Address at the Armed Forces Staff College, Norfolk, Virginia, "The Military Aspects of the Cold War," 14 November 1961.

10 (1) Roger Hilsman, "Vietnam: The Decisions to Intervene," in Jonathan R. Adelman, ed., *Superpower and Revolution* (New York: Praeger, 1986), pp. 112–14. (2) Kaufmann, p. 268.

11 (1) Vanderpool, "Oral History," p. 178. (2) Galvin, p. 275.

12 Kaufmann, pp. 58, 70, 79, 99, 150, 316.

13 Aviation Staff Officer, Special Warfare Center, Memorandum to Commanding General, SPWAR Cen, Asst. Commandant, SPWAR School, ACofS, G-3, SPWAR Center, "Reports Concerning Aviation Activities in Republic of Vietnam," 4 December 1963.

14 DoD, *Annual Reports, FY 62*, p. 100.

15 General George H. Decker, Chief of Staff, Statement Before the Subcommittee on Department of Defense Appropriations of the Committee on Appropriations, House of Representatives, 5 February 1962.

16 (1) Hilsman, pp. 114, 120. (2) Kaufmann, pp. 261–63. (3) Mike Gravel, *The Pentagon Papers (The Senator Gravel Edition)* (Boston, n.p., n.d. [c.1971]), Vol. II, pp. 13, 18–9, 23–25, 50–51, 67, 76, 80, 84, 86, 89, 101.

17 Kaufmann, p. 264.

18 United States Air Force, Headquarters, Pacific Air Forces, Memorandum for Record, "Admiral Felt's Visit with President Diem on 29 September 1961," n.d.

19 (1) Weinert, p. 38. (2) Gravel, p. 126.

20 G. Lewy, *America in Vietnam* (New York: Oxford University Press, 1978), p. 24.

21 Hilsman, pp. 133, 137.

22 Department of the Army, Assistant Chief of Staff for Force Development, "Problems Associated with the Unprogrammed Deployment of Army Aviation Assets to Vietnam," n.d. [c.1965].

23 (1) Tolson, *Airmobility*, p. 17. (2) "Army Combat Reorganization Boosts Aircraft Requirements; 1,000 Planes, 1,200 Pilots Needed; OPO's Impact on Careers Analyzed," *ANAFJ*, 10 February 1962, p. 14.

24 (1) Oden, Address to the Aviation Writers Association, p. 3. (2) "Army's Shortages in Aircraft and Pilots Worries BGen Oden," *ANAFJ&R*, 9 June 1962, p. 12.

25 Senate, Hearings Before the Committee on Government Operations, *Planning-Programming-Budgeting, Part 2*, 90th Congress, 1st Session (Washington, DC: GPO, 1967), p. 105.

26 Enthoven and Smith, pp. 62–65, 99.

27 See for example: "Curtail Pilot Shifts, DoD Orders Army; Aviation Buildup Plans Threatened," *ANAFJ*, 8 April 1961, pp. 1, 36.

28 (1) Robert S. McNamara, *Department of Defense Cost Reduction Program, First Annual Progress Report, July 8, 1963* (Washington, DC, 1963), p. 12. (2) By 1966, the cost reduction program had achieved an annual savings of $4.5 billion dollars in Department of Defense spending, with a cumulative savings of over $14 billion over a five-year period. See Robert S. McNamara, *Department of Defense Cost Reduction Program, July 8, 1966* (Washington, DC, 1966), p. 2.

29 McNamara, *The Essence of Security*, p. 91.

30 Scrutiny of the Army aviation program was a continuous process throughout this period. See (1) General Hamilton H. Howze (retired), "Howze Board Part III: Winding Up a 'Great Show,'" *Army*, April 1974, p. 19. (2) Department of the Army, "Aviation Requirements of the Combat Structure of the Army (Employment of Army Aviation in Support of Division Operations: European Theater, Volumes I–V)," 6 June 1965, Introduction, p. viii, paragraph 1.

31 (1) Tolson, *Airmobility*, pp. 17–18. (2) United States Army Military History Institute, Senior Officers Oral History Program, Army Aviation, Interview with Brigadier General Edwin L. Powell, Jr, by Colonel Bryce R. Kramer, Colonel Ralph J. Powell (Carlisle Barracks, Pennsylvania, 18 March 1978), pp. 47–50. (3) Williams, "Oral History," tape 2, pp. 53–55. (4) United States Army Military History Institute, Senior Officers Oral History Program, Project 83-6, Harry W. O. Kinnard, Lieutenant General, US Army (Retired), Interviewed by Jacob B. Couch, Jr, Lieutenant Colonel, US Army (Carlisle Barracks, Pennsylvania, 20 January 1983), pp. 1–2.

32 Goodhand, "Oral History," pp. 65, 74–75.

33 (1) Robert S. McNamara, Secretary of Defense, Memorandum for Mr Stahr, 19 April 1962. (2) Robert S. McNamara, Secretary of Defense, Memorandum for the Secretary of the Army, "Army Aviation (U)," 19 April 1962.

34 Tolson, *Airmobility*, p. 19.

35 Enthoven and Smith, pp. 99–100.

36 (1) Office of the Assistant Secretary of Defense (Comptroller), Programming (Systems Analysis), "Review of Army Aircraft Requirements," 24 January 1962, pp. 3–4. (Document sent in conjunction with McNamara's two 19 April 1962 memorandum.) (2) Office of the Assistant Secretary of Defense (Comptroller) (Systems Analysis), "Army Tactical Mobility: Issues and Opportunities," n.d. [c.1963], section III B, "Vulnerability of Army Aircraft," and section III C, "Vulnerability of the Division Base to Air Attack," pp. 36–39.

37 "The Prospects for Army Air Mobility," *Army*, March 1963, p. 20.

38 Goldberg, *Defense Documents*, pp. 330–31.

39 (1) Office of the Assistant Secretary of Defense (Comptroller) (Systems Analysis), "Army Tactical Mobility: Issues and Opportunities," n.d. [c.1963], p. 32. (2) Kaufmann, p. 287.

40 Tolson, *Airmobility*, p. 61.

41 US Army Tactical Mobility Requirements Board, "Final Report," Fort Bragg, North Carolina, 20 August 1962, pp. 8, 35–39, 78–79. Hereafter cited as TMRB, "Final Report."

42 TMRB, "Final Report," pp. 39–50, 79.

43 General Herbert B. Powell, Headquarters, United States Continental Army Command, Memorandum to the Chief of Staff, United States Army, "Report of Army Tactical Mobility Requirements Board," 20 August 1962.

44 See "Army Airmobility," *USAAD*, March 1963, pp. 3–5.

45 Perry Poe, "How's Air Mobility?," *Army*, June 1963, p. 26.

46 "Decker to Chief of Staff, USA," *Army*, September 1960, p. 18.

47 (1) General George H. Decker, Chief of Staff, US Army, "Towards Air Mobility," *Army*, November 1960, p. 18. (2) "The 'Lift' Gap," *Army*, November 1960, pp. 22, 24.

48 General George H. Decker, Chief of Staff, Statement to the House Subcommittee on Defense Appropriations, 19 April 1961.

49 "Report on Army Aircraft and Missile Systems Presented by General Lincoln," *ANAFJ*, 18 March 1961, p. 17.

50 General George H. Decker, Chief of Staff, Statement to the House Armed Services Committee, 24 February 1961. SSM 17 percent, SAM 11 percent.

51 General George H. Decker, Chief of Staff, Statement to the Senate Armed Services Committee, 6 April 1961. 5 percent for 110 fixed wing aircraft, and 6 percent for 266 helicopters.

52 "Army Aviation Will Get New Boost," *ANAFJ*, 21 January 1961, p. 27.

53 "Army Gets OK for Pilot Boost," *ANAFJ*, 11 March 1961, p. 10.

54 General George H. Decker, "Design for Ground Combat Power," *Army*, October 1962, pp. 39–40.

55 Poe, pp. 26–27.

56 (1) General Earle G. Wheeler, "Air Mobility as the Army Command Sees It," *Army*, June 1963, pp. 29–33. (2) General Earle G. Wheeler, Chief of Staff, US Army, Addresses Membership Luncheon Attendees at 19th Annual National Forum, *AHS Newsletter*, June 1963, pp. 8–9.

57 General Earle G. Wheeler, Chief of Staff, United States Army, "Airmobility," *USAAD*, October 1963, pp. 4–5.

58 Wheeler, "Air Mobility as the Army Command Sees It," p. 33.

59 General Earle G. Wheeler, Chief of Staff, United States Army, "A Perspective on Firepower and Mobility," *USAAD*, March 1964, p. 6.

60 Cyrus R. Vance, "Air Mobility as the Army Command Sees It," *Army*, June 1963, pp. 29, 32.

61 "The Prospects for Army Air Mobility," p. 21.

62 General George H. Decker, Chief of Staff, Address at the Conference of Civilian Aides, Fort Monroe, Virginia, "The Changing Army," 4 December 1961.

63 (1) "R&D Chief Orders All-Out Guerrilla Warfare Weapons Effort; 'Thought-Provoking' Plan Calls Doctrine 'Politically Sterile,'" *ANAFJ*, 8 April 1961, pp. 1, 18. (2) US Army Military History Institute, Archives, Engineer Memoirs, Lieutenant General Arthur G. Trudeau (Carlisle Barracks, February 1986). Prepared for publication by the Office of the Chief of Engineers, Dr Paul K. Walter, Historical Division. (Interview conducted in 1971), pp. 313–15. (3) Lieutenant General Arthur G. Trudeau, Chief of Research and Development, Memorandum for Directors, Office and Division Chiefs, Office of the Chief of Research and Development, "Guerrilla Warfare," 20 March 1961.

64 General George H. Decker, Chief of Staff, Closing Remarks to the House Armed Services Committee, 24 February 1961. Hereafter cited as Decker, Closing Remarks.

65 Decker, Closing Remarks.

66 (1) Decker, "The Military Aspects of the Cold War," 8 June 1961. (2) Decker, "The Military Aspects of the Cold War," 14 November 1961. (3) General George H. Decker, Chief of Staff, Address at the Armed Forces Staff College, Norfolk, Virginia, "The Military Aspects of the Cold War," 29 March 1962.

67 General George H. Decker, Chief of Staff, Statement Before the Committee on Armed Services, House of Representatives, 29 January 1962.

68 (1) Weinert, pp. 77–78. (2) Major Marquis D. Hilbert, Headquarters, United States Army Special Warfare Center, Office of the Aviation Staff Officer, Fort Bragg, North Carolina, "Comments and Observations Concerning 22nd Aviation Detachment (Special Forces) Participation as Unit of the Red Counterinsurgency Force in Joint Exercise SWIFT STRIKE III," 24 September 1963.

69 Krepinevich, pp. 109, 112–27.

70 See DoD, *Annual Reports, FY 62*, *FY 63*, and *FY 64*, pp. 99–101, 125–27, 126–28 respectively.

71 Weinert, p. 77.

72 (1) Headquarters, United States Continental Army Command, Office of the Deputy Chief of Staff for Combat Developments, Memorandum for Chief, War Games Division, "War Gaming of Airmobile Division in an Underdeveloped Area," 18 January 1962. (2) Headquarters, United Stated Continental Army Command, "War Game Evaluation of ROAD Airmobile Division in an Underdeveloped Area," Fort Monroe, Virginia, June 1962, pp. 1–2.

73 (1) John C. Flannagan, Ernest Heilberg, and John A. Lockerd, Combat Operations Research Group (CORG), CORG Memorandum CORG-M-141, "A Preliminary Study of Helicopter Tactical Forces (U)," 30 January 1962, pp. 1–2. (2) Major General Thomas F. Van Natta, DCS/CD, Memorandum to Director, CORG, "Study Request: Preliminary Study, Evaluation of Helicopter-Mounted Tactical Forces," 26 September 1961. (See also two memoranda enclosed within.)

74 Krepinevich, pp. 44–45, 108–9.

75 TMRB, "Final Report," pp. 8, 11–16, 20–27, 30–32, 73; "Brief by the President of the Board," p. 14.

76 Brigadier General Clifton F. von Kann, Director of Army Aviation, ODCSOPS, Department of the Army, "Aviation Within the Army," *The Airman*, November 1961, p. 12.

77 Elvis J. Stahr, Jr., Secretary of the Army, "The Army's Requirement for Aviation," in "1961 Army Aviation Symposium," sponsored by the AUSA. (Publication of the symposium.)

78 "The Prospects for Air Mobility," p. 21.

79 General Earle G. Wheeler, Chief of Staff, United States Army, "Army Moves Toward Mobility," *AID*, February 1964, pp. 34–35.

80 United States Army Infantry School, "Infantry Instructors' Workshop, Report of Conference," Fort Benning, Georgia, 17–20 August 1965, Major General Robert H. York, "Commandant's Welcome."

[81] Major General Edward L. Rowny, "After the Air Mobile Tests," *Army*, May 1965, p. 39.

[82] Seymour J. Deitchman, *Limited War and American Defense Policy* (Cambridge, Massachusetts: MIT Press, 1964), pp. 169–70.

[83] Tolson, *Airmobility*, p. 16.

[84] Director of Army Aviation, Office of the Deputy Chief of Staff for Military Operations, Department of the Army, "US Army Aviation Operations in South Vietnam," 1 October 1962, pp. 4, 20. (Distribution list on p. 75.)

[85] For early examples see: (1) Captain William P. Griffin, "Army Aviation in Support of Counterguerrilla Operations," *USAAD*, September 1962, pp. 9–14. (2) Major Marquis D. Hilbert and Major Everett Murray, "Use of Army Aviation in Counterguerrilla Operations," *USAAD*, October 1962, pp. 3–9. (3) Lieutenant Colonel M. W. Sutcliffe (British LNO), "Malayan Operations," *USAAD*, October 1962, pp. 10–15. (4) Lieutenant Colonel Thomas J. Sabiston, Ret, "Army Aviation Operation in Vietnam," *USAAD*, January 1963, pp. 14–20.

[86] Tolson, *Airmobility*, p. 25.

[87] See: United States Army, Strategy and Tactics Group (STAG), "War Game of the Air Assault Division," Fort Belvoir, Virginia, 14 February 1964.

[88] United States Army, Test, Evaluation and Control Group, Project TEAM, "Final Report: Field Test Program Army Air Mobility Concept," Fort Benning, Georgia, 15 January 1965, p. 3.

[89] (1) Krepinevich, pp. 120–21. (2) Tolson, *Airmobility*, pp. 29, 41, 46, 142.

[90] United States Army Combat Developments Command, "Evaluation of the Army Tactical Mobility Concept (U)," 31 January 1964, Volume I, Part 2, "USACDC Tactical Mobility Program," Appendix 5 to TAB C, "Allocation of Projects to Responsible Agency for Revised Program," pp. 2-C-3-7 to 2-C-3-10.

[91] (1) United States Army, Combat Developments Command, Aviation Agency, "Report of Conference, Former Army Aviation Commanders in Vietnam (U), Final Report," Fort Rucker, Alabama, December 1964, Section I, Introduction, paragraph 1. (2) Department of the Army, Headquarters, Combat Developments Command, Office of the Commanding General, Fort Belvoir, Virginia, Memorandum for Record, "Discussion with Mr Gene Vidal, Chairman, Mobility Panel, Army Scientific Advisory Panel," 15 May 1964.

[92] (1) US Army Utility-Tactical Transport Helicopter Company, "Achievements of the Utility-Tactical Transport Helicopter Company Supported by the Attached 571st Transportation Detachment (Aircraft Maintenance) for the Period of 1 April 1962 to 31 March 1963," n.d. [c.1963]. (2) Tolson, *Airmobility*, pp. 29–33.

[93] Brigadier General Delk M. Oden, Director of Army Aviation, Office of the Deputy Chief of Staff for Military Operations, "The Army and Air Mobility," *MR*, October 1962, p. 53.

[94] "Fire Support Helicopter Comments Asked," *Aviation Week & Space Technology*, 29 June 1964, p. 18.

[95] Tolson, *Airmobility*, pp. 144–47.

[96] Gaddis, p. 248.

[97] Paul Kesaris, ed., complied by Joan Gibson, *A Guide to the War in Vietnam: Classified Histories by the National Security Council* (Frederick, Maryland, n.d.).

[98] United States Information Service, United States Embassy, London, "US Army Forms Airmobile Division," 17 June 1965.

[99] (1) "Jungle War—Now the Army Takes to the Air," *US News & World Report*, 3 May 1965. (2) "Cavalry of the Air," *Economist*, 24 July 1965. (3) Hanson W. Baldwin, "New Role for Army," *New York Times*, 23 June 1965. (4) "US Plans Air Assault Force to Fight from Own Planes," *New York Herald Tribune*, 17 June 1965.

[100] (1) Helen Delich Bentley, "1st Cavalry to Go by Sea," *Baltimore Sun*, 4 August 1965. (2) Richard Beeston," US to send 'sky cavalry' to Vietnam," *Daily Telegraph*, 27 August 1965.

[101] Tolson, *Airmobility*, pp. 73–83.

SOURCES FOR TABLES

7-1. Bureau of Budget, *FY 64* to *FY 69, Appendix*, pp. 273; 267; 295; 309; 293–94; 294 respectively.

7-2. (1) Bureau of Budget, *FY 62–64, FY 66–69*, pp. 486; 62; 67; 70; 76; 79; 84 respectively. (2) DoD, *Semiannual Reports, Jan.–June, 1960*, p. 6. (3) DoD, *Annual Reports, FY 62–68*, pp. 366; 318; 368; 22, 382; 21–22, 434; 22, 454; 28, 498 respectively. Some figures differ between sources.

7-3. TMRB, "Final Report," p. 9.

8

EPILOGUE

EXPLAINING THE DEVELOPMENT OF AIR MOBILITY

Army High Command Support

Army aviation achieved significant progress towards air mobility in doctrinal, organizational, and material terms before 1962. The top military hierarchy understood the significance of air mobility and acted accordingly. They were not adverse to radical aviation ideas. By comparison, civilian defense officials played a minor role. Indeed, one can argue that civilian intervention in 1942, to establish organic aviation for the Field Artillery, was more important than the intervention in 1962. Scientists and engineers from the military-industrial complex played little or no role at all.

The support of successive Army Secretaries and Chiefs of Staff was a decisive factor in many Army aviation achievements. Their pro-aviation tendency, along with other top Army officials (DCSOPS and CRD), was critical in many areas. For example, their support was unmistakably present in ambitious long-term force structure planning and procurement budgeting. While budgetary reductions during the fifties affected the aviation force size, it had little effect on the direction of air mobility policy. Indeed, high-level support existed to give air mobility high priority, which ameliorated the direct effects of budget cuts.

This support may have stemmed from either military or political reasons. On the one hand, due to declining prestige after Korea, the Army Chiefs of Staff decided that the Army needed to present itself as a progressive and modern force. This may have led the chiefs to devote substantial funds to Army aviation, then seen as a glamour item with public appeal. On the other hand, the Army might have been using air mobility to redress organizational and military problems associated with new conditions of war. Glamour and prestige may have been a factor, but its importance was secondary to that of battlefield needs.

The Army high command also played a key role in supporting the doctrinal development of air mobility. Successive Army Chiefs of Staff became the main voice in stating the rationale for more air mobility. They also had an important part to play to get Department of Defense approval to procure fixed wing transport airplanes as substitutes for transport helicopters. The Army Secretariat also participated in the policy issue of primary training, because this involved the Air Force. This high-level support was crucial in issues related to the Air Force and the Department of Defense. In short, high-level attention was usually present when it involved important issues, or when making decisions on general policy guidelines.

However, at the detailed policy-making level their involvement was not always present. They sometimes delegated technical issues or some complex organizational issues to subordinate agencies to sort out among themselves. This included the issue of depot supply and maintenance, and the creation of an aviation branch. Their powers also had limits when dealing with some internal issues such as individual unit training and indoctrination. They did not involve themselves deeply in the technicalities of aircraft and equipment, or research and development policy. Nevertheless, their involvement on other more important issues was sufficient to account for the continuous progress in air mobility.

Environmental Factors

Apart from the active role played by the Army high command, environmental factors played a key role in determining the shape of air mobility. These environmental factors may include military, political, or strategic considerations. These factors provided air mobility developments with a framework, impetus and rationale. In purely military and theoretical terms, an innovation may occur because of a real or potential shortfall in military performance. Alternatively, the military may anticipate future modes of military operations radically different from the past, or perceive a military advantage in adopting a new development. The Marine Corps anticipated that future foes would possess nuclear weapons. This was a potential threat to their traditional role of amphibious assault operations. They adopted the still-immature helicopter as a future amphibious assault vehicle. Changing circumstances also led the Army down the same path.

The Army initially developed air mobility for logistical and troop movements because of its experience of poor ground transport infrastructure and the rugged terrain in Korea. The dispersion necessary on a projected nuclear battlefield then meant that ground units would often require air mobility. The emergence of guerrilla warfare generated a need for permanent air mobility for all those ground units engaged in the field. Aviation capabilities therefore became an increasingly important part of the Army.

Political and strategic developments also had an important influence on air mobility. The United States government's foreign and defense policies had to reflect new realities in the international security structure. The postwar division of the world into two superpower blocs meant that allies were a permanent feature of foreign and defense policies. It reduced the need for the United States to maintain large conventional forces of its own. This affected aviation force structure plans indirectly, by affecting the number of ground units supported. On the other hand, there was a need to aid allies at far-flung trouble spots during crises. Strategic air mobile forces were central to the desire for a quick response where there were no permanent Army forces stationed nearby.

Other strategic realities also acted as a stimulus. In the post-Korea period there was a political need to cut military expenditure to a sustainable level. This led to a reliance on tactical nuclear weapons for ground combat. Air mobility was to help the Army survive and win on the nuclear battlefield. The military understood the implications and acted accordingly without waiting for an external prompt. Later, there was a political determination to counter the rise of insurgencies around the world. Civilian defense officials acted in 1962 to accelerate air mobility development because of this political and strategic necessity. The urgency of the Vietnam situation forced the administration to send Army aviation units for operational duties before air mobile counterinsurgency concepts had fully developed. It also became the decisive factor in forming the world's first air mobile division. McNamara approved the creation of the first air mobile division because President Johnson wanted direct military intervention to halt the deteriorating situation in Vietnam. The rationales for more air mobility therefore closely reflected or integrated extremely well with administration policies and the strategic realities.

Similarly, geography and geopolitics were also important factors. The global nature of the Army's military problems coupled with a rich, productive society with a relatively small manpower base determined the necessity for strategic and tactical air mobility. The Army needed air mobility to fight numerically superior opponents. This would not only offset its weakness, but also exploit its industrial strength. Moreover, the mostly likely venues for conflict had rugged terrain, primitive surface transport infrastructure, dense vegetation, or a combination of all these. Air mobility was indispensable in these circumstances.

The introduction of helicopters and air mobility developments in foreign countries, however, was only a minor impetus to American developments. Intelligence and threat analysis on the USSR played little or no role in the process, while learning from the French and British proxy wars had a more important influence.

The Impact of Wars

Air mobility came under the influence of World War II and the Korean war during its early development and the growing Vietnam involvement during the final stages of evolution. These real combat experiences had the effect of accelerating the development process. The exigencies of war acted as a very important catalyst in the extension of Army aviation roles. Army aircraft performed roles during wartime never intended before that. The vast potential of new equipment did not become apparent until users got their hands on it and found out for themselves by trial and error. This trial and error process seemed to occur more prominently in war.

During World War II the Army used the Field Artillery's organic airplanes not just for observation missions, but also for emergency transportation of supplies and personnel. During Korea, when the majority of Army helicopters in service were the observation type, the Army similarly converted them to perform casualty evacuation and emergency resupply. Moreover, Korea had a cataclysmic effect on bringing air mobility into the Army mainstream. The Korean war also supplied important lessons for developing sky cavalry. During the sixties, Army helicopter operations in Vietnam provided positive feedback for developing air assault doctrine in counterinsurgency. The increasing Vietnam involvement served as a backdrop to the Howze Board, the divisional air mobility test, and the final decision to establish an air mobile division. Operations in Vietnam also firmly established the armed helicopter as an Army weapon.

The primary prerequisite for this functional growth was for Army aircraft to prove themselves useful and indispensable to ground commanders and staff officers. Bureaucratically Army aviation survived World War II and the postwar reorganization because its original justification at the time of establishment still held true. The conditions of war nullified Air Force attempts to limit the gradual extensions of Army aviation. The Army discarded artificial restrictions in combat in favor of, or in the interest of, expediency and minimizing casualties. Once a capability was available, no matter how humble, human ingenuity seemed inevitably to exploit this when exigencies demanded. The subsequent developments and increase in importance of Army aviation were attributable to this impetus.

Air Mobility Development Characteristics

Army aviation made its most outstanding functional gains during wartime, as a result of expediency. The salient feature during peacetime was the ability of Army aviation to defend itself bureaucratically while still making and consolidating gains. As a result Army aviation and air mobility evolved from a humble beginning ultimately into major military significance. This process involved a slow and gradual synthesis spanning over two decades.

It is arguable in theoretical terms that proponents of innovations must sell an innovation in appropriate doses. If a proposal is too radical, the authorities will dismiss it as unrealistic. If a proposal is too conservative, it becomes unexciting. Whether intentional or not, air mobility developed in small, achievable steps. Such smaller steps played an important part in developing the essential building blocks. This involved a continuous evolution of air mobility doctrine, improving the technical means of execution, and adapting the administrative and organizational structure to fit the need. By 1961 the Army was on the threshold of a major innovation, with all the prerequisites in place to support such a move. From early 1963 onward, the Army began a step-by-step build-up, training, and testing of the 11th Air Assault Division, starting from a battalion nucleus. This division then acted as a nucleus for the 1st Cavalry Division (Air Mobile). There were therefore numerous smaller steps before the final decision in 1965 to form an air mobile division. The term "creeping" extension or "incremental" growth gives the best overall description of developments in the 1942 to 1965 period. This approach was the most realistic and least risky given the circumstances.

Internal Politics

Overall, Air Force opposition was not a significant obstacle in air mobility developments. Indeed, training and indoctrination problems were more of a hindrance. Army aviation survived despite the Air Force threat of reabsorption, and flourished despite the existence of competition and restrictions. This occurred because the Army tended to be rather opportunistic, which aided the development of air mobility but was detrimental to the relationship with the Air Force. Moreover, the Army realized that its own aircraft had real combat utility in performing battle critical tasks. It therefore developed proper rationales for Army aviation that stood on their own, rather than directly competing with Air Force functions. This approach, along with the Army high command's support, negated much of the Air Force opposition. Moreover, the inadequate Air Force support of ground forces and the exigencies of war made the acquisition of a significant Army aviation capability easier. Furthermore, the Army realized in the fifties that the Secretary of Defense and Congress saw the investment in aviation as progressive. The Army exploited this fact, thereby staking a larger claim in the diminishing Army slice of the total defense budget.

Tight peacetime fiscal and executive policies also had less influence on air mobility developments than expected. The Army maintained a steady procurement policy needed to increase further the load-carrying capacity of its aircraft fleet. This was a costly inflation of requirements at a time when controlling the defense budget was a main priority in the Eisenhower administration. Aircraft research and development also remained a high priority. This state of affairs occurred because of the lax Department of Defense fiscal and policy control over the Army. Moreover, the Department of Defense tacitly supported the Army in getting the basic building blocks of air mobility. It also approved other critical developments such as sky cavalry and armed helicopters. By 1962 civilian defense officials had become the new high-level pro-active supporters of air mobility. This was important since the management of funds and programs were increasingly under their tight control.

Army Aviation's Organizational Structure

Air mobility developed despite the lack of an aviation organization. Army aviation began in 1942 as an organic part of the Field Artillery.

It remained an organization embedded in the various arms and technical services within the Army, rather than becoming a separate branch. Scholars see bureaucratic politics as an important element in military decision making. In theory no established bureaucracy would propose an innovation detrimental to itself. (An innovation is by definition a radical change in policy, involving new doctrine and force structure, and realigned procurement and budgeting.) Traditional Army elements would resist any new development that changed the internal balance of power and distribution of funds. Assigning responsibility for an innovation to an established agency should not result in satisfactory progress. Maintaining the status quo should be the norm.

The devolved aviation organization also meant that there was no powerful aviation proponent or figurehead. For example, General Howze was in a staff position, with no control over aviation units and administrative organizations. General Gavin was the Assistant Chief of Staff, G-3, and later the Chief of Research and Development. Although he may have been the intellectual force behind air mobility, he had very little day-to-day control over events. Brigadier General Carl Hutton was commandant of the Army Aviation School, with no control over policy issues. In addition, air mobility evolved over a period of 15 years. It was impossible for a single person to be in the same place to oversee its development for this length of time.

Army aviation should therefore have lacked the required bureaucratic power to protect itself from detractors and opponents. Moreover, its embedded organization was likely to generate internal disunity over aviation policies. Nevertheless, the Army did achieve significant progress in air mobility. There was little internal opposition and only some disunity over policy issues. While the Transportation Corps did tend to oppose developments that diminished its influence and prestige in Army aviation, it did not have a dominant voice, let alone a stranglehold, over decision making.

Many factors enter this equation. First, the Army Secretaries, the Chiefs of Staff, and other key officials such as the DCSOPS and CRD acted as an "inner circle" in Army aviation decision making. They were the bureaucratic power behind air mobility. The Army Chiefs of Staff were in control of all aspects of Army aviation. Collins, Ridgway, Taylor, Lemnitzer, Decker, and Wheeler acted as the figureheads. These successive Chiefs of Staff and the Army Secretaries maintained a consistent air mobility policy. Their generous support of Army aviation and active participation in decision making meant that internal disunity was never very great, and it ameliorated the Transportation Corps influence. The Army high command's determination to innovate meant that conservatism or internal hostility, if present, were minimal. It could give air mobility developments high priority and an added degree of radicalness.

There was also a degree of cooperation from the various Army elements responsible for implementation. War provided the necessary urgency in overcoming organizational handicaps. During peacetime, the progressive image of air mobility was a selling point, thereby negating the problems associated with the high cost of implementation. Elements within the Army wanted to be on the aviation bandwagon and take a slice of the "action." Thus a new military specialization may not always require a separate organization to achieve significant impact if there is high-level support for it.

Furthermore, an aviation branch or some other centralized structure may not necessarily solve the problems related to a devolved decision-making process and organizational structure. Bureaucratic infighting within the branch might have replaced interbranch bickering. Some critical organizational issues would involve difficult choices. For example, would the branch decide to become predominantly a combat arm with armed helicopters and assault forces? If so, what proportion of its aircraft would it allot to other ground elements in the support role? The specter of an aviation branch interested only in its own armed helicopter units and air mobile forces and disinterested in supporting the ground forces could therefore appear. A centralized aviation decision-making mechanism would provide no guarantees to prevent this happening. The Army preferred to maintain the generalist outlook of Army aviation and live with the unstable personnel system, rather than risk a repeat of the experience with the Air Force. As a result, the failure to establish an aviation branch probably reinforced the indoctrination in aviators' minds that supporting the ground forces was paramount. When the Army finally created an aviation branch in the early eighties, Army aviation had this point so imbued into its psyche that it had become organizational culture.

IMPLICATIONS FOR MILITARY POLICY

Policy on Technology

Technology was not very important before 1949. However, it became more important in the fifties. At first, Army aviation consisted of two-place fixed wing liaison aircraft. Even then, the Army used these aircraft in a wider variety of functions than were at first anticipated. This "growth" in aviation role was not due to improving aircraft capabilities, since aircraft with virtually the same characteristics were in use in 1945 as in 1942. The later fixed wing transport airplanes were not high technology either, but they still contributed to the functional growth of Army aviation. The same was true for the initial procurement of helicopters. The H-13 was originally for observation duties, but the Army converted it for medical evacuation in Korea. In short, the trend was first to have a growth in aviation functions, then there was the attempt to obtain more suitable or capable aircraft. This growth in Army aircraft capabilities, coupled with battlefield needs, changed the character of Army aviation completely, from a predominantly observation role to a predominantly mobility role. Technology was not important for role expansion until restrictions came into play. Forced to forego more fixed wing aircraft procurement, helicopter technology allowed the Army to pursue air mobility developments. This, however, worked both ways. When the technological limitations of larger helicopters became apparent, the Army exploited the use of fixed wing aircraft. Technological limitations also changed the aviation force structure towards more numerous but smaller transport helicopters. The key lesson here is the need for technological flexibility, to make use of existing equipment or available alternatives when necessary.

The Army's experience of using off-the-shelf equipment or civilian models demonstrated significant benefits. For example, the initial procurement of liaison aircraft in 1942 was from civilian manufacturers. The L-4 (Piper Cub) and others known as the Grasshoppers gave an excellent performance during World War II. These aircraft were better (lower cost, easier to produce) than aircraft with military specifications, such as the O-49 (L-1). The L-20 Beaver and U-1 Otter designs were originally for the Canadian wilderness. These two fixed wing transports gave sterling service, at low cost, until their replacement in the sixties. The outstanding H-13 is also noteworthy. Bell designed it with CAA certification in mind and it was the first helicopter to obtain one. The benefit of using off-the-shelf equipment is a lesson that some western militaries are relearning today. They are returning to the procurement of off-the-shelf or standard civilian models for some items. The main benefit is the lower cost with reasonable performance, while reducing procurement time.

On the other hand, dedicated military research also had its place. Transport helicopters were costly and complex machinery. Although some manufacturers produced their own earlier prototypes, subsequent funding had to come from the military services to get this technology off the ground. Moreover, the services could not avoid spending considerable sums funding turbine engine development. This was an essential technology for enabling radical air mobility concepts to materialize. Similarly, the Army and the Air Force realized very early on that the adoption of convertiplanes would solve many of their operational problems concerning airborne operations. Since commercial interests would not be enough to get this technology off the ground, military funding was essential. Had there not been continuous funding of tilt-rotor aircraft since the fifties, the V-22 would not exist today.

Roles and Missions

Air mobility is one case of a generic issue that has been an area of contention between services. The issue was the logical division of roles and missions between services. The development of STOL and VTOL aircraft technology opened up a genuine prospect of aviation support of small tactical units. During World War II logistical air support involved large troop transport aircraft dropping troops and equipment by parachute. Usually, only ground units of division size or larger received resupply by air. Division headquarters staff was usually the lowest level to coordinate close air support during World War II. Technological developments since 1945 have made combat and logistical support at battalion level or below possible and practical. It was normal for individual companies in Vietnam to receive armed and transport helicopter support. Helicopters, armed with cannons, aerial rockets, and guided missiles, gave a more intimate form of fire support. It provided more pinpoint accuracy than traditional close air support. Technology therefore opened up a vacuum between the two services.

Moreover, it was easier for the Army to operate aircraft at such devolved and intimate levels than the Air Force, which usually preferred centralization of command and operations. It was also easier for the Army to buy and operate less-sophisticated aircraft necessary for rugged operations, rather than the Air Force moving "down-market." Once having experienced performing missions with sophisticated equipment, turning to austere types was a difficult proposition.

The critical issue now is whether interservice conflicts would recur. Would new military developments (not necessarily high technology) keep making roles and missions issues materialize, and thus persist in generating friction between services? The indications are that it would not, since there is now tight hierarchical control over the services. Moreover, it seems the services have learned to accommodate each other's demands. Close air support has not been a controversial issue between the Army and the Air Force since Vietnam. In the seventies they worked out an equitable division of responsibility between fixed wing aircraft and attack helicopters. Recent developments in unmanned air vehicles have led to a joint program, with each service destined to operate their own. The post-Vietnam budgetary reductions have led to service cooperation rather than rivalry as in the post-Korea reductions. The post–Cold War reduction is also proceeding amicably. Furthermore, high technology is increasingly dominant in all services. This reduces the chance of low technology developments causing interservice friction.

Prerequisites for Innovation

Helicopter air mobility was not easy to adopt in the fifties and sixties. The high cost of transport helicopters required a large and well-funded military organization. Moreover, the complexity of helicopters meant that a highly developed industrial base was necessary. Since the United States was the most wealthy as well as the most technologically advanced country then, its military was the only organization in the world that could adopt mass air mobility. The ability to manufacture helicopters was not beyond some European countries. However, they did not possess the necessary financial resources to adopt air mobility to the extent the United States Army did. Other technologically backward societies may not even have a well-educated manpower pool to draw aviators and maintenance personnel. In general, then, a military organization must have the minimum prerequisites before it can adopt an innovation. If a military organization fails in this regard, then one can automatically assume it will fail to adopt the innovation.

It is also obvious that military innovations often include more than changes in technology. Usually there are changes associated with other aspects of preparing for war, such as the development of appropriate doctrine and organization. It may also involve new training and promotion systems.

Technology, Doctrine, and Organization

This study of air mobility points to no direct, positive relationships between technology and doctrine. The availability of technology did not automatically lead to doctrinal evolution, nor did radical doctrine mean that technological advances would automatically follow. This is a general finding already established by several scholars. When the helicopter was available during World War II, it did not lead to the development of air mobility doctrines. It is unreasonable to expect this because of the helicopter's technological infancy and limited payload capability. When the Army developed air mobility plans in the latter part of the Korean war, limited transport helicopter technology was at least available. The Army envisaged limited or occasional air mobility. In this instance technology was able to keep up with doctrine, or was in equilibrium.

The development of more ambitious Army air mobility doctrine during the Eisenhower period outstripped what was technically possible at the time. The Army high command wanted air mobility to provide Army-wide mobility in the main battle scheme. It is not evident that the existence of ambitious doctrinal plans speeded up the development of solutions to overcome the technical deficiencies of transport helicopters. This process still took its lengthy natural course. It was only after the fifties that helicopter technology caught up with doctrine again. More reliable second generation, turbine-powered helicopters made the creation of an air mobile division economical and practical. Similarly, the Army's desire to replace the parachute and assault transport did not lead to a practical convertiplane entering service.

Indeed, it is obvious from this that there was a lowest-common-denominator relationship between technology and doctrine. Without the appropriate technology, no doctrine can be effective on its own. Hence the Army found it difficult to implement the radical air mobility doctrine it developed during the fifties. This was also the case when the Marine Corps developed its vertical envelopment doctrine in the immediate post–World War II period, when helicopter technology was still in its infancy.

On the other hand, without proper doctrine, no technology can have an important influence on the conduct of military operations. This was most evident during the early stages of the Korean war, when the Army had not implemented any air mobility doctrine. By contrast, the Marine Corps had developed some advanced concepts. As a result, the Marine Corps was better able to exploit its helicopter force despite its limited capabilities. It is therefore important to have technology and doctrine in close equilibrium.

Indeed, one can include organization in this equilibrium. A dedicated sky cavalry organization, coupled with the appropriate aircraft and doctrine, would be more effective than ground reconnaissance units with ad hoc aircraft attached. Similarly, an air mobile division with integral transport helicopters would be more effective than an infantry division with helicopters assigned from higher headquarters. Technology, doctrine, and organization must be synchronized for maximum effect. Deficiencies in any one of these three areas negate advances in the other two.

One can see this lowest-common-denominator relationship in air mobility. For example, the Army initially grouped transport helicopters into homogeneous battalions and assigned them at the field army level because of their higher cost and comparatively less effective technical performance. Moreover, the doctrine was to resupply or reinforce units occasionally or only when necessary. The organization changed in 1957, when the Army introduced mixed aviation battalions because maintenance and supply problems had eased somewhat. Moreover, the Army also assigned more transport aircraft to each division because aircraft maintenance and support improved. It was also because new doctrine increasingly demanded divisions to be able to conduct their own air mobile operations using their organic transport, rather than relying on transport aircraft from higher headquarters. By 1962 transport helicopters were more capable, smaller, and had all their deficiencies solved. They therefore became easier to integrate into the division in large numbers. Forming dedicated air mobile units became practical. Integrated aviation also became more important than aviation units assigned to higher headquarters.

The organization "leg" of the triad involves not just tactical organizations. It also includes force structure, bureaucratic and administrative organizations. The Army had to modify its aviation force structure plans to take into account the production and technological problems. Army aviation's administrative structure began with a total reliance on the Air Force. As Army aviation grew larger, it slowly took them all over from the Air Force. Its bureaucratic structure and personnel system took until the eighties, when the Army created an aviation branch, before it reached a stable form.

Training is another facet of military activity associated with organization. Full-scale training of ground units in air mobile operations had to await the publication of a key field manual and for the new doctrine to filter down and spread through the Army. Higher priority for such training came only with the growing Vietnam involvement that demanded daily helicopter operations.

Developing each "leg" and integrating them will take a long implementation and maturing process. Sky cavalry provides the best example of the time needed to integrate organization, tactics, and equipment into an effective whole. The idea first appeared in 1954. It went through an iterative development process to correct the integration problems discovered during early tests and exercises. The Army initially attempted to integrate sky cavalry to ground reconnaissance units. The aircraft, ground vehicles, organization, and tactics were all unsuitable or incompatible. When the Army finally implemented a completely air-mounted air cavalry in 1961, it was only a company size unit within each division. As the Army grew more proficient in handling air cavalry organizations, such units grew in size and tactical magnitude. It ultimately evolved into air cavalry brigades for each division in the eighties.

In short, it is unrealistic to attempt assimilating a complex innovation into a military organization within too short a period. Some processes need to take their natural lengthy courses. Rather than 5 years, military organizations should allow 15 years or more for major innovations to mature.

BIBLIOGRAPHY

Research for this book's primary source material took place at three main sites in the United States: the US Army Aviation Center at Fort Rucker, Alabama; the Military History Institute at Carlisle Barracks, Pennsylvania; and Maxwell Air Force Base in Montgomery, Alabama. The Aviation Technical Library, the Aviation Museum, and the Command Historian's Office are all situated within the Army Aviation Center, while the Air University Library and the US Air Force Historical Research Agency are within Maxwell Air Force Base. These sites are archives for official military and government documents. For example, the USAF Historical Research Agency is the archive for unpublished histories of the Tactical Air Command, Air Training Command, Air Material Command, various troop carrier commands, and so on. It also possess Air Force documents relating to Army aviation.

The Military History Institute's general library, the Army Aviation Center's Technical Library, and the Air University Library are very useful for secondary sources such as books. The Air University Library and the Aviation Technical Library both also have an archives section for original military documents. Similarly, the Military History Institute's library has an archives section for Army documents and a section on old Army manuals. In addition, there is a separate Military History Institute Archives that has transcripts of oral history interviews, speech files, and personal papers of Army high-level staff officials (e.g., the Army Chiefs of Staff) and aviation officers. At the time of research, the Army Aviation Museum's material was less well organized. It generally takes laborious wild card searching and sifting for any results.

This study relies primarily on original documents for historical accuracy. Oral history interviews conducted by the Army are not a reliable source of information by comparison. This is because they occurred many years after the events. Many of the officers involved had long retired, and their memories faded.

Periodicals are an important source of material, sometimes highlighting issues not discussed even in original documents. However, because of space limitation it is impossible to list all the articles consulted for this book. The Aviation Technical Library, the Air University Library, and the Military History Institute hold complete or near-complete sets of periodical titles covering this period. Main titles include: *US Army Aviation Digest*, *Army Aviation*, *Aviation Week* (and *Space Technology*), *Army Information Digest*, *Army* (*Combat Forces Journal*), *Army Navy Air Force Journal*, *Army-Navy-Air Force Register*, *Army Navy Air Force Journal and Register*, *Army Times*, *Air Force Times*, *Military Review*, *Data*, *American Aviation*, *American Helicopter*, *National Defense Transportation Journal*, *Armor*, *Ordnance*, *Air Force*, *The National Guardsman*, *Infantry*, *Infantry School Quarterly*, and *Air University (Quarterly) Review*.

Other nonmilitary periodical titles consulted include newspapers such as the *New York Times* and *Washington Post*, and weekly publications such as *Business Week* and *US News & World Report*.

Hearings of the Appropriation and Armed Forces committees or subcommittees of both the Senate and the House of Representatives contain important details of Army aviation activities. There is also an annual review of the United States defense policies beginning 1957, prepared at the initial request of Hon. Melvin Price, Member of Congress. The author of this report was Charles H. Donnelly, Senior Specialist in National Defense from The Library of Congress, Legislative Reference Service. The title of the first report is *United States Defense Policies Since World War II*, published during the 1st Session of the 85th Congress. Subsequent reports, such as the one published in 1958, had the title *United States Defense Policies in 1957*. They contain some discussion of the Army's controversial aviation activities. All congressional documents are usually available in depository libraries.

Important documents published by government administrations and the Department of Defense include the following: the Bureau of the Budget's *The Budget of the United States Government for the Fiscal Year Ending June 30, 1946*, et cetera; and the Department of Defense's *Semiannual Report of the Secretary of Defense and the Semiannual Reports of the Secretary of the Army, Secretary of the*

Navy, Secretary of the Air Force, July 1 to December 31, 1949, et cetera. From 1 July 1958 onwards the latter became an annual report. The Government Printing Office (GPO) publishes all these congressional and official publications. Other material published by the armed forces are available through the National Technical Information Service (NTIS).

Aircraft performance data and service histories are from the annuals of *Jane's All The World's Aircraft*, *Jane's American Fighting Aircraft of the 20th Century* (London: Studio Editions, 1991) edited and compiled by Michael J. H. Taylor, as well as from F. G. Swanborough's *United States Military Aircraft Since 1908* (London: Putnam, 1971 revised edition) (with historical research by Peter M. Bowers). Richard Tierney's *The Army Aviation Story* also has a useful section of aircraft information. In addition, the 31 December 1984 (volume 33 number 12) issue of *Army Aviation* magazine also has a special article entitled "Army Aviation Equipment." For the sake of brevity, footnotes on aircraft information do not list these sources.

The names and dates in office of the Chiefs of Staff, Secretaries of War, Secretaries of the Army, and Secretaries of Defense are available in Russell Weigley's *History of the United States Army* (see below), pp. 558, 560, and Alice C. Cole, Alfred Goldberg, Samuel A. Tucker, Rudolph A. Winnacker, eds., Historical Office, Office of the Secretary of Defense, *The Department of Defense: Documents on Establishment and Organization, 1944–1978* (Washington, DC: GPO, 1978), p. 332. The latter book also contains some of the documents relating to Army aviation functions.

The following is a list of suggested reading.

PUBLICATIONS BY THE ARMED FORCES

Currey, Cecil B., Lieutenant Colonel, US Army Reserve. *With Wings as Eagles.* Washington, DC: GPO, 1984.

Department of the Army, Vietnam Studies series, Lieutenant General John J. Tolson. *Airmobility 1961–1971.* Washington, DC: GPO, 1973.

Doughty, Robert A., Major, US Army Command and General Staff College, Combat Studies Institute, Leavenworth Papers No. 1. *The Evolution of US Army Tactical Doctrine, 1946–76.* Fort Leavenworth, Kansas: GPO, August 1979.

Fails, William R., Lieutenant Colonel, Headquarters, History and Museums Division, United States Marine Corps. *Marines and Helicopters 1962–1973*. Washington, DC: GPO, 1978.

Goldberg, Alfred, and Lieutenant Colonel Donald Smith. *Army–Air Force Relations: The Close Air Support Issue (U) (A Report Prepared for*

United States Air Force Project RAND). Santa Monica, California: RAND, October 1971.

Greenfield, Kent Roberts, Colonel (Infantry, Reserve), Army Ground Forces, Historical Section. "Army Ground Forces and the Air-Ground Battle Team Including Organic Light Aviation, Study No. 35," 1948.

Hewes, James E., Jr, US Army Center of Military History. *From Root to McNamara: Army Organization and Administration, 1900–1963.* Washington, DC: GPO, 1975.

House, Jonathan M., Captain, US Army Command and General Staff College, Combat Studies Institute, Research Survey No. 2. *Toward Combined Arms Warfare: A Survey of 20th-Century Tactics, Doctrine, and Organization.* Fort Leavenworth, Kansas: GPO, August 1984.

Parker, Gary W., Lieutenant Colonel, Headquarters, History and Museums Division, United States Marine Corps. *A History of Marine Medium Helicopter Squadron 161.* Washington, DC: GPO, 1978.

Rawlins, Eugene W., Lieutenant Colonel, edited by Major William J. Sambito, Headquarters, History and Museums Division, United States Marine Corps. *Marines and Helicopters 1946–1962.* Washington, DC: GPO, 1976.

Weinert, Richard P., Historical Office, Office of the Chief of Staff, US Army Training and Doctrine Command. *History of Army Aviation 1950–1962, Phase II: 1955–1962.* Fort Monroe, Virginia, November 1976.

Weinert, Richard P., Historical Office, Office of the Deputy Chief of Staff for Military Operations and Reserve Forces, US Continental Army Command. *History of Army Aviation 1950–1962, Phase I: 1950–1954.* Fort Monroe, Virginia, June 1971.

BOOKS

Bacevich, A. J. *The Pentomic Era: The US Army Between Korea and Vietnam.* Washington, DC: National Defense University Press, 1986.

Bergerson, Frederic A. *The Army Gets an Air Force: Tactics of Insurgent Bureaucratic Politics.* Baltimore: The Johns Hopkins University Press, 1980.

Butterworth, W. E. *Flying Army: The Modern Air Arm of the US Army.* Garden City, New York: Doubleday & Co., 1971.

Coleman, J. D. *Pleiku: The Dawn of Helicopter Warfare in Vietnam.* New York: St. Martin's Press, 1988.

Francis, Devon. *Mr. Piper and His Cubs.* Ames: Iowa State University Press, n.d. [c.1973].

Galvin, John R., Colonel. *Air Assault: The Development of Airmobile Warfare.* New York: Hawthorn Books, 1969.

Krepinevich, Andrew F., Jr (Major, US Army). *The Army and Vietnam.* Baltimore: The John Hopkins University Press, 1986.

Montross, Lynn. *Cavalry of the Sky: The Story of US Marine Combat Helicopters.* New York: Harper & Brothers, 1954.

Politella, Dario. *Operation Grasshopper*. Tyler, Texas: Robert R. Longo Company, April 1958. Foreword by General Mark W. Clark.

Stanton, Shelby L. *Anatomy of a Division: The 1st Cav in Vietnam*. New York: Warner Books, 1987.

Ten Eyck, Andrew, Lieutenant Colonel. *Jeeps in the Sky: The Story of the Light Plane*. New York: Commonwealth Books, 1946.

Tierney, Richard, with editor Fred Montgomery. *The Army Aviation Story*. Northport, Alabama: Colonial Press, 1963.

Wakefield, Ken. *The Fighting Grasshoppers: US Liaison Aircraft Operations in Europe, 1942–1945*. Leicester: Midland Counties Publications, 1990.

Weigley, Russell F. *History of the United States Army*. New York: MacMillan Co., 1967.

MAJOR ARTICLES BY TOP ARMY OFFICIALS

"Army Airmobility." *US Army Aviation Digest*, Vol. 9, No. 3 (March 1963), pp. 3–5.

"Army Operations in Limited War." *Army Information Digest*, June 1958, pp. 41–48.

Brucker, Wilber M. "How Far We Have Come!" *Army*, September 1960, pp. 39–40.

Brucker, Wilber M. "Look at the Army Whole." *Army*, December 1957, pp. 26–31.

Brucker, Wilber M. "This is a Modern Army." *Army*, September 1959, pp. 27–31.

Brucker, Wilber M. "Why the Army Needs Aviation." *American Aviation*, 10 March 1958, p. 21.

Brucker, Wilber M., and General Maxwell D. Taylor. "Message." *US Army Aviation Digest*, June 1957, pp. 2–3.

Brucker, Wilber M., and General Maxwell D. Taylor. "The Army of the Future." *Army Information Digest*, January 1957, pp. 14–16.

Cairns, Bogardus S., Brigadier General. "Army Aviation = Battlefield Mobility." *Army Information Digest*, August 1958, pp. 2–14.

Collins, J. Lawton, General. "Modern Weapons for Today's Army." *Army Information Digest*, Vol. 5, No. 6 (June 1950), pp. 3–9.

Daley, John P., Major General. "Mobility Is a Must." *Ordnance*, March–April 1958, p. 797.

Decker, George H., General. "Design for Ground Combat Power." *Army*, October 1962, pp. 35–40.

Decker, George H., General. "Towards Air Mobility." *Army*, November 1960, p. 18.

Easterbrook, Ernest F., Brigadier General. "Army Aviation Looks Ahead." *US Army Aviation Digest*, Vol. 5, No. 12 (December 1959), pp. 1–4.

Easterbrook, Ernest F., Brigadier General. "Army Aviation Philosophy." *US Army Aviation Digest*, Vol. 5, No. 10. (October 1959), pp. 1–3.

Eddleman, C. D., Lieutenant General. "Men, Missiles and Atomics on the Futurarmy Battlefield." *Army*, December 1956, pp. 24–30.

Eddleman, C. D., Lieutenant General. "Military Operations." *Army Information Digest*, February 1957, pp. 16–23.

Eddleman, Clyde D., General. "Mobility and Future Land Warfare." *Army Information Digest*, May 1962, pp. 26–31.

Gavin, James M., Lieutenant General. "Freedom to Think Boldly." *Army*, September 1960, p. 72.

Gavin, James M., Lieutenant General. "New Developments in Prospect." *Army Information Digest*, January 1957, pp. 16–17.

Gavin, James M., Lieutenant General. "The Mobility Differential." *Army*, June 1963, pp. 34–36.

Gavin, James M., Major General. "Cavalry, and I Don't Mean Horses!" *Harpers Magazine*, April 1954, pp. 54–60.

Howze, Hamilton H., General. "Tactical Employment of the Air Assault Division." *Army*, September 1963, pp. 35–53.

Howze, Hamilton H., Major General. "Army Aviation Looks Ahead." *Army Information Digest*, July 1957, pp. 2–13.

Howze, Hamilton H., Major General. "Combat Tactics for Tomorrow's Army." *Army*, Vol. 8, No. 3 (October 1957), pp. 24–30.

Howze, Hamilton H., Major General. "Future of Army Aviation." *US Army Aviation Digest*, Vol. 3, No. 6 (June 1957), pp. 4–6.

Howze, Hamilton H., Major General. "Helicopters in the Army." *Ordnance*, January–February 1958, pp. 634–38.

Howze, Hamilton H., Major General. "Soldiers With a Double Skill." *Army*, January 1957, pp. 30–34.

Howze, Hamilton H., Major General. "The Army's Stake in the Helicopter." *Army Aviation*, June 1956, pp. 2, 34–35.

Howze, Hamilton H., Major General. "The Future Direction of Army Aviation." *Army*, Vol. 7, No. 5 (December 1956), pp. 51–54.

Howze, Hamilton H., Major General. "The Land Battle in an Atomic War." *Army*, July 1961, pp. 28–36.

Howze, Hamilton H., Major General, et al. "The Last Three Years of Army Aviation." *US Army Aviation Digest*, Vol. 4, No. 3 (March 1958), pp. 2–60.

Hutton, Carl I., Brigadier General. "Army Aviation in the Modern Army." *National Defense Transportation Journal*, March–April 1959, pp. 38–40, 65–67.

Hutton, Carl I., Brigadier General. "Aviation is the Answer." *Army Aviation*, July 1956, pp. 2, 38–40.

Hutton, Carl I., Brigadier General. "Future Prospects of Army Aviation." *Army Information Digest*, November 1956, pp. 18–29.

Hutton, Carl I., Brigadier General. "Helicopter-Borne Operations." *US Army Aviation Digest*, Vol. 4, No. 10 (October 1958), pp. 5–8.

Johnson, Earl D., Under Secretary of the Army. No title. *National Defense Transportation Journal*, September–October 1953, p. 64.

McNamara, Robert S., Secretary of Defense. "The Prospect for Army Air Mobility." *Army*, March 1963, pp. 20–21.

Meyer, Richard D., Brigadier General. "The Next Decade in Army Aviation." *US Army Aviation Digest*, Vol. 5, No. 8 (August 1958), pp. 27–30.

Oakes, John C., Lieutenant General. "Facets of Mobility." *Army*, September 1960, pp. 50–51.

Oden, Delk M., Brigadier General. "The Army and Air Mobility." *Military Review*, October 1962, pp. 52–53.

Pickett, George B., Jr, Brigadier General. "The Army's Tactical Mobility Concept." *US Army Aviation Digest*, Vol. 10, No. 11 (November 1964), pp. 1–5.

Powell, Herbert B., General. "Air Cavalry Troop." *Army Information Digest*, February 1962, pp. 67–74.

Powell, Herbert B., Major General. "The Infantryman in the Atomic Age." *Army Information Digest*, April 1958, pp. 32–39.

Rich, Charles W. G., Major General. "Air Assault Needs Materiel Improvement." *Army*, October 1963, pp. 16, 64.

Rich, Charles W. G., Major General. "Industry and Air Assault." *US Army Aviation Digest*, Vol. 10, No. 2 (February 1964), pp. 2–6.

Ridgway, Matthew B., General. "An Army On Its Toes." *The Army Combat Forces Journal*, December 1954, pp. 10–11.

Ridgway, Matthew B., Lieutenant General. Address Before the 8th Infantry Division Association, "Trends in Modern Warfare." 19 November 1949, in "Digest of Speeches," *Army Information Digest*, Vol. 5, No. 1 (January 1950), p. 63.

Rowny, Edward L., Major General. "After the Air Mobile Tests." *Army*, May 1965, pp. 36–39.

Taylor, Maxwell D., General. "'Safety Lies Forward'—Technologically and Tactically." *Army*, December 1956, pp. 20–24.

Taylor, Maxwell D., General. "The Army's Need for Mobility." *National Defense Transportation Journal*, November–December 1957, pp. 45, 75–78.

Taylor, Maxwell D., General. "The Changing Army." *Army*, October 1955, p. 10.

Taylor, Maxwell, General. "The Army Needs Mobility." *National Defense Transportation Journal*, November–December 1955, Vol. 11, pp. 50, 52.

Teller, Edward, Dr. "The Impact of Nuclear Weapons on Future Operations." *Army Information Digest*, January 1957, p. 18.

Trudeau, Arthur G., Lieutenant General. "Air Mobility Requirements." *Army*, June 1959, p. 58.

Trudeau, Arthur G., Lieutenant General. "Toward a Mobile Army." *National Defense Transportation Journal*, November–December 1959, pp. 37, 60–63.

Trudeau, Arthur G., Lieutenant General. "Toward a More Mobile Army." *National Defense Transportation Journal*, November–December 1958, pp. 46–47, 84–85.

Vance, Cyrus R. "Air Mobility as the Army Sees It." *Army*, June 1963, pp. 29, 32–33.

Vidal, Eugene. "Airlift All the Way to Battle." *Army*, September 1960, pp. 71–72.

von Kann, Clifton F., Brigadier General. "Army Aviation." *Army*, November 1960, pp. 40–53.

von Kann, Clifton F., Brigadier General. "Air Mobility Problems." *Army*, September 1960, pp. 53–54.

von Kann, Clifton F., Brigadier General. "Mobility Now." *Army Information Digest*, April 1960, pp. 32–43.

von Kann, Clifton F., Brigadier General. "The Aerial Reconnaissance and Security Troop." *Army Aviation*, 28 February 1960, pp. 65–66, 68.

von Kann, Clifton F., Brigadier General. "The Real Goal of Army Aviation." *US Army Aviation Digest*, Vol. 6, No. 3 (March 1960), pp. 35–37.

von Kann, Clifton F., Brigadier General. "Vulnerability of Army Aircraft." *Military Review*, Vol. 41, No. 11 (November 1961), pp. 2–8.

Wheeler, Earle G., General. "A Perspective on Firepower and Mobility." *US Army Aviation Digest*, Vol. 10, No. 3 (March 1964), pp. 2–6.

Wheeler, Earle G., General. "Air Mobility as the Army Sees It." *Army*, June 1963, pp. 29–32.

Wheeler, Earle G., General. "Airmobility." *US Army Aviation Digest*, Vol. 9, No. 10 (October 1963), pp. 2–5.

Wheeler, Earle G., General. "Army Moves Toward Mobility." *Army Information Digest*, February 1964, pp. 32–37.

Wheeler, Earle G., Major General. "Strategic Mobility." *Army Information Digest*, January 1957, pp. 2–12.

Yount, Paul F., Brigadier General. "Air Transportation and the Army." *National Defense Transportation Journal*, September–October 1953, pp. 62–63, 66–65.

Yount, Paul F., Major General, Army Chief of Transportation. "Transportation for Peace." *National Defense Transportation Journal*, July–August 1957, pp. 57–64.

Yount, Paul F., Major General. "Transportation for Tomorrow's Army." *Army Information Digest*, February 1958, pp. 12–19.

Index

About the Author

CHRISTOPHER C.S. CHENG is a Manager of the Yung Shing Enterprise Company in Hong Kong. He has a doctorate from the Department of War Studies, King's College, University of London.